Chemistry of
Nanocrystalline Oxide Materials
Combustion Synthesis, Properties and Applications

Chemistry of
Nanocrystalline Oxide Materials
Combustion Synthesis, Properties and Applications

K C Patil
M S Hegde
IISc. Bangalore, India

Tanu Rattan
Nunano Solutions, India

S T Aruna
National Aerospace Laboratories, India

 **World Scientific**

NEW JERSEY · LONDON · SINGAPORE · BEIJING · SHANGHAI · HONG KONG · TAIPEI · CHENNAI

CHEM
0166643926

Published by

World Scientific Publishing Co. Pte. Ltd.
5 Toh Tuck Link, Singapore 596224
USA office: 27 Warren Street, Suite 401-402, Hackensack, NJ 07601
UK office: 57 Shelton Street, Covent Garden, London WC2H 9HE

British Library Cataloguing-in-Publication Data
A catalogue record for this book is available from the British Library.

ISBN-13 978-981-279-314-0
ISBN-10 981-279-314-3

Typeset by Stallion Press
Email: enquiries@stallionpress.com

Printed by FuIsland Offset Printing (S) Pte Ltd, Singapore

Dedicated to our spouses

| Prabha | Malati | Sanjay | Mahesh |
| (KCP) | (MSH) | (Tanu) | (STA) |

Contents

Foreword

Nanoscience and technology has emerged as a frontier area of research today. A great deal of stimulating progress is being made in the world of nanoscience every other day. Research in "Nanochemistry," chemistry of nanomaterials, has resulted in the developments of newer methods of synthesis of materials with desired structure, composition, and properties of nanomaterials and their related applications. With this rich array of knowledge available, the question arises as to how does a student, a researcher, or materials scientist get an actual feel of nanopowders and get down to the science and art of making them?

Chemistry deals with the synthesis and analysis both being complementary are equally important. Prof KC Patil with his training in Inorganic Synthesis and Propellant Chemistry has found a novel and simple method of tailor-making oxide materials. Prof Patil's method is distinctly different from the other known methods. I am particularly impressed by the simplicity of the process and the choice of the reactants to prepare a variety of oxide materials from alumina to zirconia and their composites.

Self-propagating high-temperature synthesis where simple chemical processes are used to study complex chemical reactions is of interest from both theoretical and practical point of view. The solution combustion method of Prof Patil is of interest from the investigations of the mechanism of the physico-chemical processes involved, dynamics of the product formation, their stability limit, and control of the process. These studies will no doubt bring out new and important results and better understanding.I trust the contributions and the achievements described in the book will stimulate further investigations in the area of Nanoscience and Technology.

I wish this book all success it deserves.

Prof AG Merzhanov
Academician of Russian Academy of Sciences
Chief Scientific Adviser
Institute of Structural Macrokinetics and Materials Science (ISMAN)
Chernogolovka, Moscow reg. 142432, Russia

Preface

Developing new routes to nanocrystalline materials is a challenging task for solid-state chemists and materials scientists. The book on *Chemistry of Nanocrystalline Oxide Materials* (*Combustion Synthesis, Properties, and Applications*) is the result of the research work carried out by the authors and their PhD students during the last two and half decades. This book introduces an innovative pathway to the synthesis of oxide materials. The interest in nano-oxides is due to their technological applications in fields like microelectronics, catalysis, coatings, energy storage, and environment protection and remediation. Synthesis of nano-oxides along with their characterization, physicochemical properties, and applications is of interest to students, teachers, researchers, and materials scientists alike.

Chemists carrying out reactions with ions in solution have been practicing nanochemistry for centuries. The technique of making nanocrystalline oxide materials by "combustion process" is an integrated approach, combining both breaking-down and building-up processes of producing nanomaterials. "Catalysis" is a classical example of "nanotechnology." The solution combustion synthesis method has already made inroads into nanoscience and nanotechnology but never before it has been documented and made available to researchers working in this field. This book is an endeavor to comprehensively put together all the published work of the authors in one treatise. This method of making oxide materials being simple and fast is being practiced by many individuals and groups around the world. No attempt is made here to quote the work done by others.

The authors would like to thank Dr(s) KC Adiga, R Soundar Rajan, Jayant Budkuley, M Ramanath, JJ Vittal, D Gajapathy, C Nesamani,

P Ravindranthan, S Govindarajan, NRS Kumar, JJ Kingsley, S Sundar Manoharan, K Suresh, R Gopichandran, MMA Sekar, N Arul Dhas, S Ekamabaram, M Muthuraman, S Samrat Ghosh, T Sushil Kumar Rajan, Parthasarathi Bera, Arup Gayen, K Nagaveni, Tinku Baidya, AS Prakash, Sounak Roy, Sudhanshu Sharma, Asha Gupta, B Angadi, and B Nagappa whose contributions have helped made this book possible.

Prof Patil gratefully acknowledges the facilities provided by the Department of Inorganic and Physical Chemistry, Indian Institute of Science, Bangalore.

We appreciate the support and cooperation of Prof Arun Umarji and Dr GT Chandrappa while writing the book. Sincere thanks are due to Mr Sanjay Rattan and Dr Madhavi for their invaluable contributions in editing and proof reading the book from conception to completion.

We thank the American Chemical Society for granting us the permission to reproduce some of the figures as such. The authors would like to thank World Scientific editors especially Lim Sook Cheng and Lee Kok Leong, for all their hard work and patience in correcting and editing the manuscript.

Finally, sincere and grateful thanks are due to Prof AG Merzhanov who readily agreed to write the foreword and has been a source of constant support and encouragement.

We hope the book will serve students and researchers working in the field of Materials Science and Engineering and NanoChemistry.

KC Patil
TanuRattan
ST Aruna
MS Hegde

Chapter 1

Introduction

1.1 GENERAL

The study of metal oxides has attracted the attention of materials scientists due to their optical, electrical, magnetic, mechanical, and catalytic properties, which make them technologically useful. Ferromagnetic iron oxides, γ-Fe_2O_3, Fe_3O_4, spinels, and hexaferrites are materials of choice for data storage and transmission. Ferroelectric and dielectric oxides like $BaTiO_3$, $Pb(Zr,Ti)O_3$ with perovskite structure are extensively used in electronic devices. Physical properties of perovskite oxides such as electrical, electronic, magnetic, and optical vary with composition. For example, $LaNiO_3$ is a metallic oxide while $LaMnO_3$ is an antiferromagnetic insulating oxide. Partial substitution of La ion by Sr, Ca, Ba, or Pb, makes it metallic as well as ferromagnetic. The discovery of superconductivity in $La_{1.85}Sr_{0.15}CuO_4$ and $YBa_2Cu_3O_{7-\delta}$, popularly known as 1-2-3 compound, with superconducting property at 90 K spurred interest in the chemistry of oxide materials. The relation between the structure and properties (both physical and chemical) of oxide materials and their applications are of great importance.[1]

Some important oxide materials and their applications have been summarized in Table 1.1.

Currently considerable interest in nanocrystalline oxide materials exists owing to their unusual properties. Decreasing particle size results in some remarkable phenomenon. It has been found that smaller the particles, the

- Higher the catalytic activity (Pt/Al_2O_3).
- Higher the mechanical reinforcement (carbon black in rubber).
- Higher the electrical conductivity of ceramics (CeO_2).

- Lower the electrical conductivity of metals (Cu, Ni, Fe, Co, and Cu alloys).
- Higher the photocatalytic activity (TiO_2).
- Higher the luminescence of semiconductors.
- Higher the blue-shift of optical spectra of quantum dots.
- Higher the hardness and strength of metals and alloys.
- Superparamagnetic behavior of magnetic oxides.

Table 1.1.   Properties and applications of oxide materials.

Oxides	Property	Applications
Al_2O_3, CeO_2	Hardness	Abrasive
TiO_2, CeO_2, Fe_2O_3	Catalysts	Air and water purification
M^o/Al_2O_3 (M^o = Cu, Ag, Au, Pt, and Pd) $Ce_{1-x}M_xO_{2-\delta}$ (M = Cu, Ag, Au, Pt, Pd, Rh, and Ru)	Redox catalyst	Three-way catalyst for automobile exhausts
$Ti_{1-x}M_xO_{2-\delta}$ (M = Cu, Ag, Pt, Pd, Mn, Fe, Co, Ni and W)	Photocatalyst	Oxidation of organic matter
TiO_2, ZnO	UV–Vis sunlight absorbing	Photocatalyst, sunscreen, and paint
MTi/ZrO_3 (M = Ca, Sr, Ba, and Pb); PZT	Dielectric	Sensors, MEMS
γ-Fe_2O_3, $BaFe_{12}O_{19}$, MFe_2O_4	Super paramagnetic	Cancer detection and remediation, sensors and memory devices
TiO_2, Fe_2O_3, Cr_2O_3, MAl_2O_4, MCr_2O_4 (M = transition metal ions)	Colors	Ceramic pigments
M/Al_2O_3 (M = Cr^{3+}, Co^{2+}, Ni^{2+})		
M/ZrO_2, $RE/M:ZrSiO_4$ (M = Fe^{3+}, Mn^{2+}, V^{4+}; RE = rare earth ion)		
Eu^{3+}/Y_2O_3 (red), Eu^{2+}, Tb^{3+}/Ba-hexaaluminate	Luminescence	Phosphors-CFL, color TV picture tube
Al_2O_3, ZrO_2, ZTA, mullite, cordierite, tialite	Refractory	Toughened ceramics
MgO, CaO, and ZnO	Adsorbent	Defluoridation and COD from paper mill effluents
$YSZ(Y_2O_3$–$ZrO_2)$	Electrolyte	Solid oxide fuel cell materials
Ni/YSZ	Anode	
$La(Sr)MO_3$, M = Mn, Cr	Cathode/interconnect	

Unusual optical and electrical properties in these materials take place due to a phenomenon known as quantum confinement. The large surface area to volume ratio of nanomaterials leads to their use as catalysts. Excellent sintering characteristics of these fine powders are useful in ceramics and composites. Dispersion of minute particles in various fluids allows the fabrication of corrosion resistant coatings and thin films.

1.2 PREPARATIVE METHODS

One of the challenges faced by materials scientists today is the synthesis of materials with desired composition, structure, and properties for specific applications. While one can evolve a well-reasoned approach to the synthesis of oxide materials, serendipity has played an important role in making new materials. Rational synthesis of materials require knowledge of crystal chemistry, besides thermodynamics, phase equilibrium, and reaction kinetics. The physicochemical properties of many materials are determined by the choice of synthetic methods. Selection of the synthetic route is crucial to control the composition, structure, and morphology of a chosen material. For instance, barium hexaferrite ($BaFe_{12}O_{19}$) can be used either as a permanent magnet or as a recording media depending on the morphology of the compound, which in turn is dependent on the preparation route.

Oxide materials are usually prepared by solid-state reactions, i.e., either by the ceramic method or by precipitation from solution and subsequent decomposition. A variety of metal oxides, both simple and complex, are prepared by the conventional ceramic method. This involves the mixing of constituent metal oxides, carbonates, etc., and their repeated heating and grinding. These methods are used on both laboratory and industrial scale. However, there is an increasing demand for alternate routes to the synthesis of oxide materials that give superior properties when compared with those available from conventional methods. It should not be construed that conventional methods are substandard in any way; they are still used in the industrial production of several oxide materials. Nonetheless, the need for alternate synthesis routes for oxide materials has arisen because of intrinsic problems relating to:

1. Inhomogeneity of the products obtained by ceramic methods.
2. Incorporation of chemical impurities during repeated grinding and heating operations. Impurities have a deleterious effect on the high-temperature

mechanical behavior of engineering ceramics and on the electrical properties of electroceramics.

3. Coarseness of particles obtained from conventional routes, which make them unsuitable for coatings.

Nonuniform powder compositions make reproducible component fabrication difficult because of chemical inhomogeneity and voids in microstructure. Greater purity and homogeneity from novel methods can lead to improved physical properties.

The present trend is to avoid brute force methods in order to have a better control of stoichiometry, structure, and phase purity of metal oxides. Soft chemical routes are now increasingly becoming important to prepare a variety of oxides including nanocrystalline oxide materials. These approaches make use of simple chemical reactions like coprecipitation, sol–gel, ion exchange, hydrolysis, acid leaching, and so on, at considerably low temperatures compared to the ceramic method. Use of precursors, intercalation reactions, electrochemical methods, hydrothermal process, and self-propagating high-temperature synthesis (SHS) are some of the other contemporary methods. Several books and review articles on the synthesis of oxide materials have been published over the years.[2–12] Among these methods, combustion or fire synthesis (SHS) is quite simple, fast, and economical.

Although SHS has been successfully used to make non-oxide materials, its application for synthesis of oxide materials was delayed due to economic reasons. Furthermore, it being a solid-state method, phase purity and particle size control is not possible. Also, due to its high-temperature course it is not suitable for the preparation of nanocrystalline materials. In this context, a low-temperature initiated combustion method[13,14] developed by Patil *et al.* at Indian Institute of Science, Bangalore has carved a niche. This low-temperature initiated self-propagating combustion process is different from the well-known Pechini (citrate process) which uses external heating at high temperatures to burn away the extra carbon.[10]

Combustion process is different from pyrolysis since once ignited it does not require external heating. In the synthesis of nanomaterials by soft routes there are two approaches: (i) breaking-down and (ii) building-up processes. Solution combustion synthesis of nanocrystalline oxide materials while appearing to be a breaking-down process is in fact an integrated approach, as the desired oxide products nucleate and grow from the combustion residue.

1.3 SCOPE OF THE BOOK

Future technologies based on nanoscale inorganic solids will hinge on the production of a controlled size, shape, and structure of nanocrystalline oxide material to make a working device. Today, solution combustion synthesis of oxide materials is becoming popular and is being used by materials scientists all over the world.[13–17] This process is particularly useful for preparing such nanocrystalline oxides. Yet, research and development of the process and its application to nano-oxide materials has not caught the full attention of Nanoscientists and Engineers. The objective of this book is to comprehensively present the solution combustion process and give ready recipes to prepare a wide range of oxide materials. It is gratifying to note that the preparation of nanophosphors[18] and aluminas[19] by the solution combustion method has been described as simple laboratory experiment for undergraduate students.

This book contains 10 chapters. The introductory section has already highlighted the importance of the chemistry of oxide materials, their properties and applications, and the need for a novel method for the preparation of nanocrystalline oxide materials.

Chapter 2 describes the preparation of nanocrystalline oxide materials by the combustion of solid precursors. The precursors discussed are: metal hydrazine carboxylates like formate, acetate, oxalate, and hydrazine carboxylate and their solid solutions. Unlike the precursors described in literature which decompose at high temperatures, these complexes ignite at low temperatures and undergo autocatalytic combustion with the evolution of large amount of gases to yield voluminous nanocrystalline oxides. Several nanocrystalline metal oxides prepared by the combustible solid precursor method are useful as catalysts, magnetic, and dielectric materials.

Chapter 3 introduces the reader to the unique solution combustion synthesis route of making oxide materials with desired composition, structure, and properties for specific applications. Combustion being an exothermic redox chemical reaction requires an oxidizer and a fuel. The oxidizers used are water soluble metal nitrates; the fuels employed are readily available compounds like urea, glycine, metal acetates, and hydrazides, all of which are also water soluble. Synthesis of metal oxides is achieved by rapidly heating an aqueous solution containing stoichiometric quantities of the redox mixture. Calculation of stoichiometry is very critical and important to solution combustion synthesis. This calculation is based on the principles used in propellant chemistry and consists of balancing elemental oxidizing and reducing valences of the compounds utilized in combustion. The chapter deals with the theory of combustion reaction and its thermodynamic calculation. Typical procedures for preparation of desired oxides are given. The role of fuels and the conditions for synthesis of the nanocrystalline oxide materials are discussed. Advantages of the solution combustion method over other methods are highlighted.

Chapter 4 deals with the synthesis of technologically important alumina and related oxide materials. Urea appears to be an ideal fuel for the combustion synthesis of α-Al_2O_3, metal aluminates, and metal-ion-doped aluminas. Nanocrystalline aluminas are also prepared by using glycine and metal acetate precursors. Luminescent materials like Cr^{3+}-doped alumina, aluminates, and garnets, as well as Ce^{3+}-, Tb^{3+}-, and Eu^{2+}-doped hexaaluminates have been prepared and investigated. The synthesis of cobalt (Co^{2+})-and chromium (Cr^{3+})-doped blue and pink aluminas as ceramic pigments are examined. Synthesis and properties of Al_2O_3–SiO_2 and Al_2O_3–TiO_2 composites are also presented.

Chapter 5 gives an account of the preparation and catalytic properties of nanocrystalline CeO_2, $Ce_{1-x}(Zr/Ti)_xO_2$, and metal-ion-substituted ceria ($Ce_{1-x}M_xO_{2-\delta}$ where M = Cu, Ag, Au, Pt, Pd, Rh, and Ru). Platinum-ion-substituted ceria and palladium-ion-substituted ceria are discovered as catalysts for H_2–O_2 recombination and three-way catalytic converter for automobile exhaust, respectively. Oxalic acid dihydrazide (ODH) is the preferred fuel to obtain nanocrystalline ceria.

Chapter 6 contains the preparation and properties of nanocrystalline magnetic oxides based on Fe_2O_3. The materials prepared are simple ferrites (MFe_2O_4), mixed (Li–Zn, Mg–Zn, and Ni–Zn) ferrites, orthoferrites ($LnFeO_3$), garnets, and hexaferrites. ODH appears to be an ideal fuel for the preparation of iron-containing nanocrystalline magnetic oxides, since combustion is flameless or smoldering type.

Chapter 7 discusses the synthesis and photocatalytic properties of nanocrystalline anatase titania (TiO_2). Preparation and photocatalytic properties of metal-ion-substituted TiO_2 are investigated for the degradation of organic dyes and compared with anatase titania and metal-impregnated titania. Titanate-based minerals have been prepared and studied as Synroc materials for nuclear waste immobilization. Both, tetraformyl trisazine and glycine are suitable for combustion synthesis of titania and titania-based oxide materials.

Synthesis of t-, m-, c-ZrO_2, and related oxide materials is the topic of Chap. 8. The oxides studied are stabilized forms of ZrO_2, rare earth ion doped zirconia and zircon pigments, zirconia–ceria (ZrO_2–CeO_2), zirconia–titania (ZrO_2–TiO_2), and pyrochlores (ZrO_2–Ln_2O_3) systems. Synthesis of NASICONs has also been reviewed. Carbohydrazide is the chosen fuel for the preparation of zirconia and related oxides.

Chapter 9 describes the synthesis, properties, and applications of perovskite oxides (ABO_3) — metal titanates ($MTiO_3$), zirconates ($MZrO_3$), $LnMO_3$, and strontium-substituted perovskites ($Ln(Sr)MO_3$, where M = Cr, Mn, Fe, Co). Dielectric oxide materials like $MTiO_3$, $MZrO_3$ (where M = Ca, Ba, Sr, and Pb), PZT, and lead niobates have also been prepared and their properties are studied.

The synthesis and properties of technologically important oxides not covered in the earlier chapters constitutes the subject matter of Chap. 10. The oxides prepared include simple metal oxides, metal borates, silicates, and vanadates. The properties of combustion synthesized metal chromites, rare earth oxide based pigments, and high-T_c materials are also examined.

An 'Appendix' giving details regarding the preparation of hydrazine-based fuels used for solution combustion synthesis of oxide materials as well as the oxidizer titanyl nitrate is provided. Also some important tips for the preparation of oxides using heterogeneous solutions are given.

It is hoped that this book will serve as a handbook on solution combustion synthesis of oxide materials and will prove to be useful to Chemists, Physicists, Materials Scientists, and Students in this field. The information on the synthesis of nanocrystalline oxide materials is a significant contribution to the field of nanoscience and nanotechnology and in the final analysis is expected to inspire entrepreneurship.

References

1. Mc Carrol WH, Ramanujachary KV, Oxides: Solid-state chemistry, in King RB (ed.), *Encyclopedia of Inorganic Chemistry* 2nd ed., Wiley, pp. 4006–4053, 2005.
2. Ozin GA, Arsenault AC, *Nanochemistry: A Chemical Approach to Nanomaterials*, Royal Society of Chemistry, London, 2005.

3. Rao CNR, Muller A, Cheetham AK (eds.), *The Chemistry of Nanomaterial Synthesis, Properties and Applications*, 2 Vol, Wiley-VCH, Weinheim, Germany, 2004.

4. Cao G, *Nanostructures and Nanomaterials, Synthesis, Properties and Applications*, Imperial College Press, London, 2004.

5. Rao CNR, *Chemical Approaches to the Synthesis of Inorganic Materials*, Wiley Eastern Ltd., New Age International Ltd., New Delhi, 1994.

6. Hagenmuller P (ed.), *Preparative Methods in Solid State Chemistry*, Academic Press, New York, 1992.

7. David Segal, *Chemical Synthesis of Advanced Ceramic Materials*, Cambridge Univ. Press, Cambridge, 1989.

8. Jolivet JP, Henry M, Livage J, *Metal Oxide Chemistry and Synthesis: From Solution to Solid State*, Wiley, New York, 2000.

9. Vallet-Regi M, Preparative strategies for controlling structure and morphology of metal oxides, in Rao KJ (ed.), *Perspectives of Solid State Chemistry*, Narosa, Publishing House, New Delhi, pp. 37–65, 1995.

10. Segal D, Chemical synthesis of ceramic material, *J Mater Chem* 7: 1297–1305, 1997.

11. Klabunde KJ, Mohs C, Nanoparticles and nanostructural materials, in Interrante LV, Hamplan-Smith MJ (eds.), *Chemistry of Advanced Materials: An Overview*, Wiley Inc, pp. 271–327, 1998.

12. Nersesyan MD, Merzhanov AG, SHS of complex oxides, in Borisov AA, De Luca L, Merzhanov AG (eds.), *Self-propagating High-temperature Synthesis of Materials*, Taylor and Francis, New York, pp. 176–187, 2002.

13. Patil KC, Aruna ST, Ekambaram S, Combustion synthesis, *Curr Opin Solid State Mater Sci* 2: 158–165, 1997.

14. Patil KC, Aruna ST, Mimani T, Combustion synthesis: An update, *Curr Opin Solid State Mater Sci* 6: 507–512, 2002.

15. Merzhanov AG, Theory and practice of SHS: Worldwide state of the art and the newest results, *Int J Self-Propagating High-Temp Synth* 2 (2): 113–158, 1993.

16. Varma A, Diakov V, Shafirovich E, Heterogeneous combustion: Recent developments and new opportunities for chemical engineers, *AIChE J* 51: 2876–2884, 2005.

17. Segadaes AM, Oxide powder synthesis by the combustion route, *Euro Ceram News Lett* 9: 1–5, 2006.

18. Bolstad DB, Diaz AL, Synthesis and characterization of nanocrystalline $Y_2O_3:Eu^{3+}$ phosphor, *J Chem Ed.* 79: 1101–1104, 2002.

19. Tanu Mimani, Fire synthesis: Preparation of alumina products, *Resonance* 5: 50–57, 2000.

Chapter 2

Combustible Solid Precursors to Nanocrystalline Oxide Materials

2.1 INTRODUCTION

Simple metal oxides are usually prepared by thermal decomposition of metal salts like carbonates, oxalates, nitrates, acetates, etc. On the other hand, mixed metal oxides like ferrites, chromites, and manganites are obtained from single source precursors containing more than one metal ion in the desired mole ratio in the same solid matrix.[1] A large number of technologically important mixed metal oxides such as ferrites, chromites, manganites, and titanates have been prepared by the decomposition of the corresponding mixed metal oxalates, carbonates, cyanides, or hydroxides. Barium titanate ($BaTiO_3$) for example, is an important ferroelectric oxide prepared by the thermal decomposition of $BaTiO(C_2O_4)_2$.

$$BaTiO(C_2O_4)_2 \xrightarrow[\sim 700°C]{O_2} BaTiO_3 + 4CO_2 \qquad (1)$$

Generally, the temperature required for decomposition is in the range of 700°C to 1000°C (Table 2.1).

Because of high calcination temperatures and long duration of heating, the oxide products obtained from these precursors are coarse. To make fine-particle oxide materials it is desirable to have precursors that decompose exothermically at low temperatures with evolution of large amounts of gases. The combustion of a redox compound like ammonium dichromate yielding voluminous green chromium oxide is a good example of this type of decomposition (Fig. 2.1 — an artificial volcano).[8] Ammonium dichromate having both oxidizing and

Table 2.1.   Single source precursors to metal oxides.

Single source precursors	Decomposition temperature, $T_d(^\circ C)$	Products	References
$(NH_4)_2M(CrO_4)_2 \cdot 6H_2O$, M = Mg, Ni, ...	1000–1100	MCr_2O_4	2
$MCr_2O_7 \cdot 4C_2H_5N$, M = Mn, Co	1200	MCr_2O_4	2
$LiCr(C_2O_4)_2$	1000	$LiCrO_2$	2
$M_3Fe_6(CH_3COO)_{17}O_3$ $OH \cdot 12C_5H_5N$	1000–1300	MFe_2O_4	2
$LaFe_{0.5}Co_{0.5}(CN)_6 \cdot 5H_2O$	800	$LaFe_{0.5}Co_{0.5}O_3$	3
$LaFe(CN)_6 \cdot 6H_2O$	800	$LaFeO_3$	3
$LaCo(CN)_6 \cdot 5H_2O$	800	$LaCoO_3$	3
$Ln_{1-x}M_x(OH)_3$	600	$LnMO_3$	3
$Ca_{2-x-y}M_xM'_yCO_3$	930–1080	$Ca_2M_xM'_yO_n$	4
$MTiO(C_2O_4)_2 \cdot 4H_2O$, M = Ca, Sr, Ba, and Pb	700–1000	$MTiO_3$	5
$(NH_4)_3Al_{1-x}Fe_x$ $(C_2O_4)_3 \cdot nH_2O$	800	Fe/Al_2O_3	6
$YBa_2Cu_3(N_2O_2)_{2.6}OH_xO_y$	650–700	$YBa_2Cu_3O_{6+\delta}$	7

Fig. 2.1.   Artificial volcano (green product Cr_2O_3).

reducing groups undergoes self-sustained combustion reaction once ignited due to the oxidation of ammonia by the dichromate, as shown below:

$$(NH_4)_2Cr_2O_7(s) \xrightarrow{\Delta} Cr_2O_3(s) + 4H_2O(g) + N_2(g) \qquad (2)$$

In the context of precursors that decompose exothermically to make fine-particle oxide materials, the chemistry of metal hydrazine and metal hydrazine carboxylate precursors is of relevance.

2.2 COMBUSTIBLE METAL HYDRAZINE AND METAL HYDRAZINE CARBOXYLATE COMPLEXES

Conventionally, hydrazine (N_2H_4) is used as a mono propellant and rocket fuel. The high-energy required for the propulsion of the rocket comes from the decomposition of hydrazine to nitrogen, ammonia, and hydrogen:

$$2N_2H_4 \longrightarrow N_2 + 2NH_3 + H_2 + \text{Heat} \qquad (3)$$

In this respect, the chemistry of hydrazine (H_2N-NH_2) and its derivatives is interesting.[9,10] Structure and properties of hydrazine is summarized below.

Physical properties

m.p. $= 3.16°C$, b.p. $= 114.37°C$
Density $= 1.004\,\mathrm{g\,cm^{-3}}$
Dielectric constant $= 57.7$
Dipole moment $= 1.83\text{--}1.85\,\mathrm{D}$
$\Delta H_{\text{formation}} = +50.434\,\mathrm{kJ\,mol^{-1}}$ (Endo)
$\Delta H_{\text{decomposition}} = -148.64\,\mathrm{kJ\,mol^{-1}}$ (Exo)

Hydrazine (Fig. 2.2) has positive heat of formation ($\Delta H_f^\circ = {\sim}50.434\,\mathrm{kJ\,mol^{-1}}$) and is therefore thermodynamically unstable.

It has two free lone pairs of electrons, four substitutable H-atoms, and more interestingly, a potent high-energy N–N bond. The presence of lone electron pairs allows hydrazine to coordinate to a metal ion either as a unidentate or bridged bidentate ligand (Fig. 2.3) to form complexes.

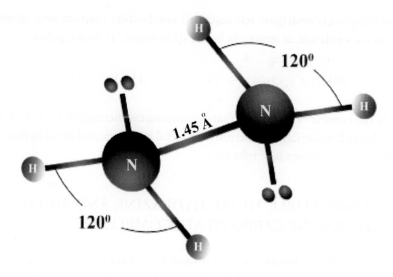

Fig. 2.2.   The structure of hydrazine.

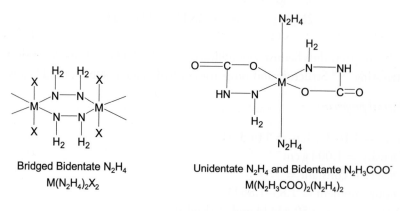

Bridged Bidentate N_2H_4
$M(N_2H_4)_2X_2$

Unidentate N_2H_4 and Bidentante $N_2H_3COO^-$
$M(N_2H_3COO)_2(N_2H_4)_2$

Fig. 2.3.   Coordination of hydrazine (N_2H_4) and $N_2H_3COO^-$.

Hydrazine has great affinity for CO_2 and forms hydrazine car-boxylic acid (carbazic acid), N_2H_3COOH.[11] The hydrazine carboxylate anion $N_2H_3COO^-$ is a bidentate ligand and several metal hydrazine carboxylate complexes like $M(N_2H_3COO)_2$, $M(N_2H_3COO)_2 \cdot nH_2O$, $M(N_2H_3COO)_2(N_2H_4)_2$, or $N_2H_5M(N_2H_3COO)_3H_2O$ are known.[12,13]

Several metal hydrazine complexes have been prepared and investigated as low-temperature precursors to oxide materials.[14] The various complexes prepared are shown in Fig. 2.4.

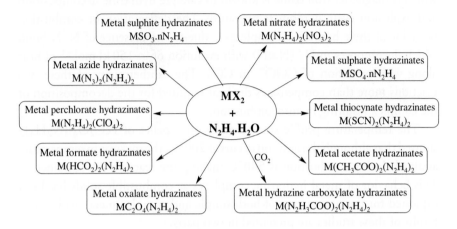

Fig. 2.4.   Metal hydrazines and metal hydrazine carboxylate complexes.

Interestingly, it was observed that the thermal reactivity of these complexes dramatically changes from detonation to deflagration (controlled combustion) to decomposition (Table 2.2).[15]

Table 2.2.   Thermal reactivity of metal hydrazine complexes.

$M(N_2H_4)_n^{2+}$	$ClO_4^-, NO_3^-,$ N_3^-	$N_2H_3COO^-, C_2O_4^{2-}$ $HCOO^-, CH_3COO^-$	SO_4^{2-}, SCN^-, X^- (X = halogens)
M = (Cr, Mn, Fe, Co, Ni, Zn, Cu, Cd)	Detonation	Deflagration (precursor to fine oxide)	Decomposition
M = (Mg, Al)	Deflagration (precursor to fine oxide)		

For example, complexes containing perchlorate, nitrate, and azide anions of transition metals explode or detonate,[16] whereas those of nontransition metals, e.g., Mg and Al deflagrate. Hydrazine complexes with anions like sulfate and halides simply decompose with loss of hydrazine leaving behind metal sulfates and halides. Metal hydrazine complexes with carboxylate anions also deflagrate. Both $FeC_2O_4(N_2H_4)_2$ and $Fe(N_2H_3COO)_2(N_2H_4)_2$ when ignited at ~200°C, burn (combust) to yield voluminous iron oxide (γ-Fe_2O_3) which is magnetic. Iron oxide is known to catalyze hydrazine decomposition and explosion even at room temperature. The deflagration or combustion behavior of these hydrazine complexes is due to the presence of N–N bond which decomposes to N_2 (N≡N) with evolution of ~150 kJ mol^{-1} of heat along with oxidation of COO^- to CO_2. The exothermicity of these two reactions more than compensates the heat required for the decomposition of the complex resulting in autocombustion.

The temperature profile studies made on a pellet of $FeC_2O_4 \cdot (N_2H_4)_2$ showed that although the ignition occurs at 200°C, the maximum temperature attained during combustion is 540°C making the reaction self-sustained.[17] This behavior of metal hydrazine complexes as combustible solids has been exploited for the preparation of both simple and complex metal oxides. The results of these studies are presented in two parts.

Part I: Metal hydrazine carboxylates: Precursors to simple metal oxides.

Part II: Single source precursors to mixed metal oxides.

PART I: METAL HYDRAZINE CARBOXYLATES: PRECURSORS TO SIMPLE METAL OXIDES

2.3 PREPARATION OF METAL FORMATE, ACETATE, OXALATE, AND HYDRAZINE CARBOXYLATES

These complexes can be prepared by the addition of hydrazine hydrate to an aqueous solution of the metal salts. An instantaneous reaction takes place with the evolution of heat. The preparation of metal formate hydrazinates — $M(HCOO)_2(N_2H_4)_2$, metal acetate hydrazinates — $M(CH_3COO)_2(N_2H_4)_2$, metal oxalate hydrazinates — $MC_2O_4(N_2H_4)_2$, where M = Mn, Co, Ni, Zn, and Cd, involves the addition of excess hydrazine hydrate (99–100%) to the corresponding metal formate[18] or acetate[19] or oxalate hydrates.[20]

Typical equations for the formation of various complexes are given below:

Formate:

$$M(HCOO)_2 \cdot (H_2O)_2 + 2N_2H_4 \cdot H_2O \longrightarrow M(HCOO)_2(N_2H_4)_2 + 4H_2O \tag{4}$$

Acetate:

$$M(CH_3COO)_2 \cdot xH_2O + 2N_2H_4 \cdot H_2O$$
$$\longrightarrow M(CH_3COO)_2(N_2H_4)_2 + (x+2)H_2O \tag{5}$$

Oxalate:

$$MC_2O_4 \cdot xH_2O + 2N_2H_4 \cdot H_2O \longrightarrow MC_2O_4(N_2H_4)_2 + (x+2)H_2O \tag{6}$$

The reaction of metal powder or aqueous solutions of metal salts with ammonium oxalate in hydrazine hydrate also give the oxalate hydrazinates.

$$M^\circ + (NH_4)_2C_2O_4 + 2N_2H_4 \cdot H_2O$$
$$\longrightarrow MC_2O_4(N_2H_4)_2 + 2H_2O + H_2 + 2NH_3 \tag{7}$$

Metal hydrazine carboxylates can be prepared by the addition of hydrazine hydrate to an aqueous solution of the metal salts and then saturating the solution with CO_2 gas or dry ice. It is known that CO_2 gets incorporated into N_2H_4 forming $N_2H_3COO^-$ anion, which being a bidentate ligand, coordinates with the metal ion.

$$N_2H_3COOH + N_2H_4 \cdot H_2O \longrightarrow N_2H_5COON_2H_3 + H_2O \tag{8}$$

Three types of complexes form depending upon the concentration of the ligands and the metal, as seen below.[16–18]:

$$MX_2(aq) + 2N_2H_5COON_2H_3 + 2H_2O$$
$$\longrightarrow M(N_2H_3COO)_2 \cdot 2H_2O + 2N_2H_5X$$
$$(M = Mg, Ca, Mn, TiO, and ZrO) \tag{9}$$

$$LnX_3(aq) + 3N_2H_5COON_2H_3 + 3H_2O$$
$$\longrightarrow Ln(N_2H_3COO)_3 \cdot 3H_2O + 3N_2H_5X$$
$$(Ln = rare earth ion) \tag{10}$$

$$MX_2(aq) + 2N_2H_5COON_2H_3 + 2N_2H_4 \cdot H_2O$$
$$\longrightarrow M(N_2H_3COO)_2(N_2H_4)_2 + 2N_2H_5X + 2H_2O$$
$$(M = Fe, Co, Ni, \text{ and } Zn) \tag{11}$$

$$MX_2(aq) + 3N_2H_5COON_2H_3 + H_2O$$
$$\longrightarrow N_2H_5M(N_2H_3COO)_3 \cdot H_2O + 2N_2H_5X$$
$$(M = Mg, Mn, Fe, Co, Ni, Zn, \text{ and } Cd) \tag{12}$$

The formation and characterization of all these complexes have been studied thoroughly.[12,13,21,22]

2.3.1 *Thermal Analysis and Combustion of Metal Hydrazine Carboxylates*

Thermal properties of metal hydrazine carboxylates have been investigated using TG–DTA. Typical thermograms representing each class of carboxylates are shown in Figs. 2.5–2.10.[11–18]

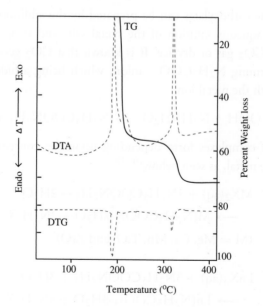

Fig. 2.5.   TG–DTA–DTG curves of $Mn(CH_3COO)_2(N_2H_4)_2$.

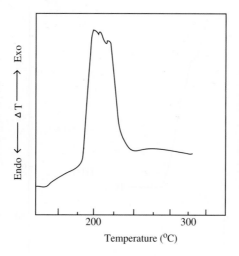

Fig. 2.6. DTA of $Fe(C_2O_4)(N_2H_4)_2$.

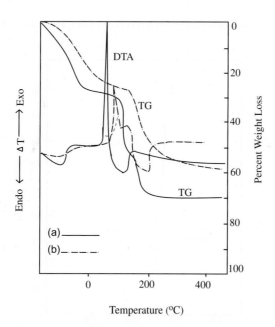

Fig. 2.7. TG–DTA curves of (a) $TiO(N_2H_3COO)_2 \cdot 2H_2O$ and (b) $ZrO(N_2H_3COO)_2 \cdot 2H_2O$.

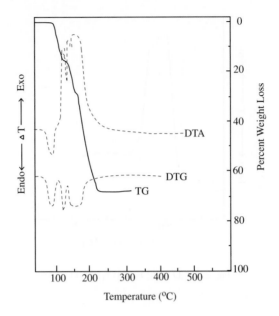

Fig. 2.8.   Simultaneous TG–DTA–DTG of Mn(N₂H₃COO)₂(H₂O)₂.

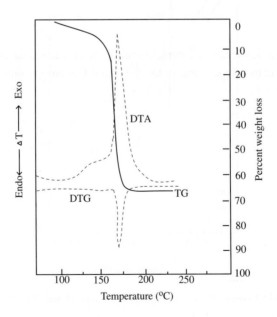

Fig. 2.9.   Simultaneous TG–DTA–DTG of Fe(N₂H₃COO)₂(N₂H₄)₂.

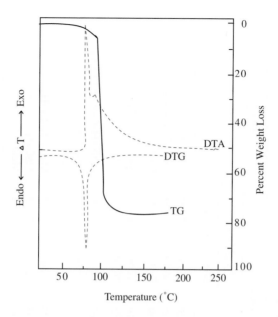

Fig. 2.10.   Simultaneous TG–DTA–DTG of $N_2H_5Fe(N_2H_3COO)_3 \cdot H_2O$.

Thermogravimetry of all metal hydrazine carboxylates shows multiple step decomposition to yield the related oxides. The intermediate steps have been assigned to the formation of corresponding metal carboxylates, carbonates, and oxy-carbonates depending upon the metal (transition metal, nontransition metal, or rare earth metal). However, hydrazinium metal hydrazino carboxylate complexes decompose through an oxalate hydrazinate intermediate (Table 2.3).

The DTA of the carboxylates initially shows endothermic peaks due to the loss of monodentate hydrazine at low temperatures, e.g., formates, whereas loss of bidentate or bridged hydrazine, e.g., oxalates that decompose at high temperature, are exothermic in nature. Metal hydrazine complexes like formates, oxalates, and hydrazine carboxylates all decompose to their respective oxides at considerably low temperatures compared to their parent salts due to the synergistic oxidation reaction of both hydrazine and carboxylate ions. This has been confirmed from the data shown on the exothermic DTA peak temperatures in Table 2.4.

There is a distinct difference between thermal decomposition and combustion; the former is a slow process with a heating rate of $10°$/min, whereas in the

Table 2.3.   Thermal analysis of hydrazinium metal hydrazino carboxylates.

Compound	DTA peak Temp. (°C)	Thermogravimetry				Products
		Step no.	Temp. range (°C)	Weight loss (%)		
				Observed	Required	
$N_2H_5Fe(N_2H_3COO)_3 \cdot H_2O$	165,180 (exo)	1	145–220	76.00	75.93	$\gamma\text{-}Fe_2O_3$
$N_2H_5Co(N_2H_3COO)_3 \cdot H_2O$	175 (exo)	1	170–245	36.50	37.61	$CoC_2O_4 \cdot 2N_2H_4$
	245 (exo)	2	245–280	76.10	76.03	Co_3O_4
$N_2H_5Ni(N_2H_3COO)_3 \cdot H_2O$	182 (exo)	1	170–210	48.00	46.60	$NiC_2O_4 \cdot N_2H_4$
	193,220 (exo)	2	210–385	76.00	77.67	NiO
$N_2H_5Zn(N_2H_3COO)_3 \cdot H_2O$	145 (endo)	1	70–165	18.00	18.16	$Zn(N_2H_3COO)_2 (N_2H_4)_2$
	195 (exo)	2	165–215	53.00	55.07	ZnC_2O_4
	240 (exo)	3	215–390	76.00	76.17	ZnO
$N_2H_5Mg(N_2H_3COO)_3$	120 (endo)	1	50–135	36.00	37.52	$MgC_2O_4 \cdot 2N_2H_4$
	320 (exo)	2	135–343	72.20	70.13	$MgCO_3$
	390 (endo)	3	343–418	86.00	85.72	MgO

Table 2.4.   DTA peak (exothermic) temperatures (°C) of metal hydrazine carboxylates.

Carboxylates	Cr	Mn	Fe	Co	Ni	Cu	Zn
Formate (COO$^-$)	—	272	—	200–280	217	—	—
Acetate (CH$_3$COO$^-$)	—	150–275	—	180–305	205–330	—	—
Oxalate (C$_2$O$_4^{2-}$)	—	217	202	209–262	219–290	—	—
Hydrazine carboxylate N$_2$H$_3$COO$^-$	190	175–195	135–175	190–200	195–228	115	240

latter, the heating rate is fast (100°/min). All the metal hydrazine carboxylate complexes like iron oxalate hydrazinate and iron hydrazine carboxylate hydrazinate when heated rapidly undergo a combustion reaction. These complexes once ignited ($\sim$200°C) undergo autocatalytic decomposition and combustion (burn) using atmospheric oxygen. They evolve large amounts of gases like CO_2, H_2O, NH_3, N_2, etc. and yield voluminous combustion residue (ash). The reaction is self-sustained due to the presence of hydrazine moieties such as $N_2H_3^-$, N_2H_4, or $N_2H_5^-$ along with COO$^-$ groups that undergo oxidation in the presence of atmospheric oxygen. The products of combustion are often crystalline and of nanosize with large surface area. Theoretical equations for the combustion reaction of metal hydrazine carboxylate precursors are given below.

$$M(N_2H_3COO)_2(s) \xrightarrow[\text{Air}]{1/2O_2} MO(s) + 2CO_2(g) + 2NH_3(g) + N_2(g) \quad (13)$$

$$M(N_2H_3COO)_2(N_2H_4)_2(s) \xrightarrow[\text{Air}]{O_2} MO(s) + 2CO_2(g) + 4NH_3(g)$$
$$+ H_2O(g) + 2N_2(g) \quad (14)$$

$$2M(N_2H_3COO)_2(N_2H_4)_2(s) \xrightarrow[\text{Air}]{2.5O_2} M_2O_3(s) + 4CO_2(g) + 8NH_3(g)$$
$$+ 2H_2O(g) + 2N_2(g) \quad (15)$$

$$N_2H_5M(N_2H_3COO)_3(H_2O)(s) \xrightarrow[\text{Air}]{O_2} MO(s) + 3CO_2(g) + 4NH_3(g)$$
$$+ 2H_2O(g) + 2N_2(g) \quad (16)$$

$$2N_2H_5M(N_2H_3COO)_3(H_2O)(s) \xrightarrow[\text{Air}]{O_2} M_2O_3(s) + 6CO_2(g) + 8NH_3(g)$$
$$+ 4H_2O(g) + 4N_2(g) \qquad (17)$$

An important result of this observation has led to the preparation of the most widely used recording material γ-Fe$_2$O$_3$, in a single step, by the decomposition of iron hydrazine carboxylate precursors. The normal method of synthesizing γ-Fe$_2$O$_3$ involves three steps:

  (i) Precipitation of α-FeOOH,
 (ii) Dehydration and reduction of α-FeOOH to Fe$_3$O$_4$, and
(iii) Oxidation of Fe$_3$O$_4$ to γ-Fe$_2$O$_3$.

A critical step during the formation of γ-Fe$_2$O$_3$ is the oxidation of Fe$_3$O$_4$. This, in turn is controlled by the particle size of Fe$_3$O$_4$ and its surface area. A smaller size of the particles not only facilitates trapping of moisture but also controls the rate of diffusion of oxygen. Thus, it appears that the preparation of γ-Fe$_2$O$_3$ by the normal method is longer and more tedious. Although both N$_2$H$_5$Fe(N$_2$H$_3$COO)$_3$·H$_2$O and Fe(N$_2$H$_3$COO)$_2$(N$_2$H$_4$)$_2$ complexes decompose exothermically at $\sim$200°C, the bluish green crystal (Fig. 2.11a) of these precursors when ignited with a hot filament or match stick, burn without flame. The precursors undergo linear combustion (layer by layer) to give ribbon like red iron oxide (Pharoah's snake) (Fig. 2.11b) instantaneously.

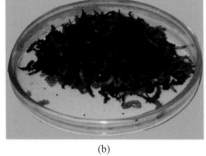

(a)   (b)

Fig. 2.11.   (a) Bluish green crystals of N$_2$H$_5$Fe(N$_2$H$_3$COO)$_3$·H$_2$O and (b) reddish brown iron oxide.

Theoretical equations for the combustion reaction can be represented as follows:

$$2Fe(N_2H_3COO)_2(N_2H_4)_2(s)$$

$$\xrightarrow[\text{Air}]{2.5\,O_2} Fe_2O_3(s) + 4CO_2(g) + 8NH_3(g) + 2H_2O(g) + 4N_2(g)$$

$$\text{(18 mol of gases/mol of Fe}_2\text{O}_3) \quad (18)$$

$$2N_2H_5Fe(N_2H_3COO)_3(H_2O)(s)$$

$$\xrightarrow[\text{Air}]{2.5\,O_2} Fe_2O_3(s) + 6CO_2(g) + 8NH_3(g) + 4H_2O(g) + 4N_2(g)$$

$$\text{(22 mol of gases/mol of Fe}_2\text{O}_3) \quad (19)$$

The red oxide was identified as γ-Fe_2O_3 by its characteristic powder X-ray diffraction pattern (Fig. 2.12). The DTA of the red oxide showed an exothermic peak at 595°C corresponding to the conversion of γ-Fe_2O_3 to α-Fe_2O_3.[23]

The BET surface area of γ-Fe_2O_3 obtained from the thermal decomposition of the complexes is typically between 40 and 70 $m^2\,g^{-1}$ and the particle

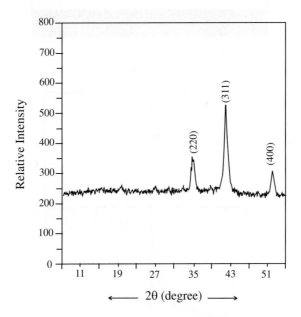

Fig. 2.12. XRD pattern of γ-Fe_2O_3.

size is in the nano-range as shown by TEM (Fig. 2.13). Because of the fine nature of these particles the *M–H* curve shows a narrow hysteresis curve indicating superparamagnetic nature of the oxide (Fig. 2.14). The Mössbauer spectrum of the residue recorded at room temperature shows hyperfine splitting pattern characteristic of γ-Fe_2O_3 (Fig. 2.15).

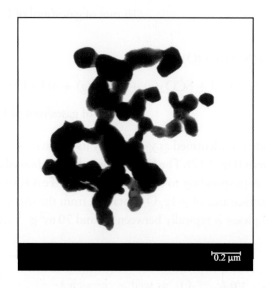

Fig. 2.13.   TEM of γ-Fe_2O_3.

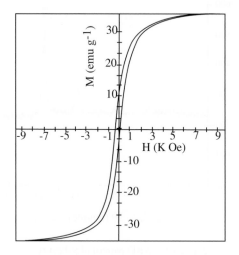

Fig. 2.14.   *M–H* curve of γ-Fe_2O_3.

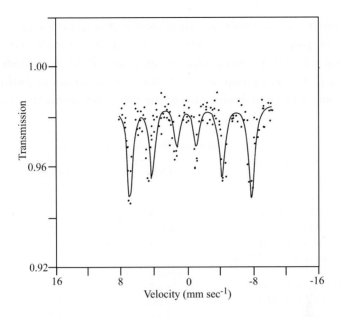

Fig. 2.15. Mössbauer spectrum of γ-Fe$_2$O$_3$.

The particulate and magnetic properties of γ-Fe$_2$O$_3$ (I) prepared by the combustion of both the precursors (Fe(N$_2$H$_3$COO)$_2$·(N$_2$H$_4$)$_2$ and N$_2$H$_5$Fe(N$_2$H$_3$COO)$_3$·H$_2$O), as well as through Fe$_3$O$_4$, i.e., γ-Fe$_2$O$_3$ (II) are summarized in Table 2.5.

Table 2.5. Particulate and magnetic properties of γ-Fe$_2$O$_3$.

Red oxide		Lattice constant a (nm)	Crystallite size (nm)	Surface area (m^2 g^{-1})	Coercivity (Oe)
γ-Fe$_2$O$_3$(I)	A	0.837	18	38	180
	B	0.835	13	75	130
γ-Fe$_2$O$_3$(II)	A	0.837	16	31	144
	B	0.836	10	65	120

A — Fe(N$_2$H$_3$COO)$_2$·(N$_2$H$_4$)$_2$; B — N$_2$H$_5$Fe(N$_2$H$_3$COO)$_3$·H$_2$O.

γ-Fe$_2$O$_3$ prepared by the combustion of precursor B has larger surface area and smaller particle size compared to the one prepared by the combustion of precursor A due to the evolution of more gases as seen from the combustion reactions (Eqs. (18) and (19)).

This interesting thermal reactivity of metal hydrazine carboxylates has been used for the preparation of a variety of simple and doped oxide materials like Co-doped γ-Fe_2O_3 (recording material) and $Ce_{1-x}Pr_xO_2$ (red pigment) and $Y_{0.975}Eu_{0.025}O_3$ (red phosphor).[24-26] Some of the oxides prepared by the combustion of metal hydrazine carboxylate precursors are listed in Table 2.6.

Table 2.6.   Technologically useful oxide materials.

Precursors	Oxide	Surface area $(m^2\,g^{-1})$	Particle size (nm)
$Fe(N_2H_3COO)_2\cdot(N_2H_4)_2$	γ-Fe_2O_3	68	28
$Fe(N_2H_3COO)_2\cdot(N_2H_4)_2$	$Fe_3O_4^a$	49	42
$N_2H_5Fe(N_2H_3COO)_3\cdot H_2O$	γ-Fe_2O_3	75	27
$N_2H_5Co_xFe_{1-x}(N_2H_3COO)_3$ $\cdot H_2O$	Co-doped γ-Fe_2O_3	75	27
$N_2H_5Fe(N_2H_3COO)_3\cdot H_2O$	$Fe_3O_4{}^a$	71	29
$Zn(N_2H_3COO)_2\cdot(N_2H_4)_2$	ZnO	67	27
$TiO(N_2H_3COO)_2\cdot 2H_2O$	TiO_2	114	22
$ZrO(N_2H_3COO)_2\cdot 2H_2O$	ZrO_2	14	121
$Ce(N_2H_3COO)_3\cdot 3H_2O$	CeO_2	90	15
$Pr_xCe_{1-x}(N_2H_3COO)_3\cdot 3H_2O$	Pr-doped CeO_2	90	15
$Y(N_2H_3COO)_3\cdot 3H_2O$	Y_2O_3	55	36
$Eu_xY_{1-x}(N_2H_3COO)_3\cdot 3H_2O$	Eu^{3+}-doped Y_2O_3	55	36

[a] Obtained from combustion in vacuum.

PART II: SINGLE SOURCE PRECURSORS TO MIXED METAL OXIDES

2.4 MIXED METAL OXIDES

Mixed metal oxides like cobaltites (MCo_2O_4), ferrites (MFe_2O_4), manganites (MMn_2O_4), and titanates ($MTiO_3$) have been prepared by the combustion of corresponding mixed metal acetate, oxalate, and hydrazine carboxylate precursors. Since the ionic radii of 3d transition elements are similar, i.e., $Mn^{2+} = 0.083\,nm$, $Fe^{2+} = 0.078\,nm$ and $Co^{2+} = 0.0745\,nm$ and the crystal structures of metal oxalate and hydrazine carboxylates being isomorphous, solid solutions are formed by them in the entire range.

2.4.1 *Mixed Metal Acetate and Oxalate Hydrazinates: Precursors to Cobaltites*

The above method has been used to prepare mixed metal acetate and oxalate hydrazinates $M_{1/3}Co_{2/3}(CH_3COO)_2(N_2H_4)_2$ (where $M = Ni^{2+}$ or Zn^{2+}); $NiCo_2(C_2O_4)_3(N_2H_4)_6$ and $MgCo_2(C_2O_4)_3(N_2H_4)_5$ as precursors of nickel, zinc, and magnesium cobaltites. The acetate complexes were made by the addition of an excess of alcoholic hydrazine hydrate to a solution containing a mixture of the corresponding metal acetates, in the molar ratio of 1:2. The precipitated complexes were filtered, washed with ethanol and diethyl ether and stored in a vacuum desiccator over P_2O_5. These precursors decomposed exothermically in the temperature range of 165–345°C to yield the corresponding cobaltites.[27]

The oxalate complexes were made by dissolving the corresponding metal powders in a solution of ammonium oxalate and excess hydrazine hydrate.[28]

$$Mg^0 + 2Co^0 + 3(NH_4)_2C_2O_4(aq) + 5N_2H_4 \cdot H_2O$$
$$\longrightarrow MgCo_2(C_2O_4)_3(N_2H_4)_5 + 6NH_3 + 8H_2O + 3H_2 \quad (20)$$

The formation of spinel cobaltites is confirmed by powder XRD pattern (Fig. 2.16). The average crystallite sizes of $NiCo_2O_4$ and $ZnCo_2O_4$ calculated from X-ray line broadening using the Scherrer equation are 10 and 16 nm, respectively.

2.4.2 *Mixed Metal Oxalate Hydrazinates: Precursors to Spinel Ferrites*

Mixed metal oxalate hydrazinates are prepared either from metal powders or metal salts. Stoichiometric quantities of the respective metal powders are dissolved in a solution of ammonium oxalate in hydrazine hydrate. After the dissolution of the metal powders, alcohol is added to precipitate out the complex. It is then washed with alcohol and ether.

$$M^0 + 2Fe^0 + 3(NH_4)_2C_2O_4(aq) + 6N_2H_4 \cdot H_2O$$
$$\longrightarrow MFe_2(C_2O_4)_3(N_2H_4)_6 + 6NH_3 + 9H_2O + 3H_2 \quad (21)$$

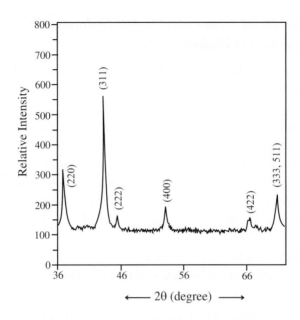

Fig. 2.16.   XRD pattern of $ZnCo_2O_4$ (Co K_α).

Mixed metal oxalate hydrazinates $MFe_2(C_2O_4)_3(N_2H_4)_x$ (M = Mg, Mn, Co, Ni, Zn, and $x = 5$ or 6) can also be obtained by the reaction of mixed metal oxalate hydrates $MFe_2(C_2O_4)_3(H_2O)_6$ with excess of hydrazine hydrate.

The resulting complexes are crystalline solids with characteristic colors. As the reactions are carried out under reducing conditions iron is present as Fe^{2+} in these complexes. The presence of Fe^{2+} makes them rather susceptible to atmospheric oxidation and therefore they need to be prepared and stored in an atmosphere of nitrogen. They are unstable and on storage decompose by losing hydrazine.

All these complexes undergo a single-step decomposition forming the corresponding ferrites (MFe_2O_4). Only in the case of the magnesium complex, a second exothermic peak is observed which has been attributed to the recrystallization of $MgFe_2O_4$. Thus, mixed metal ferrites are formed at temperatures as low as 150–160°C. The mixed metal oxalate hydrates do not yield ferrites at such low temperatures, e.g., magnesium oxalate hexahydrate, gives a mixture of α-Fe_2O_3, MgO, and magnesium ferrite ($MgFe_2O_4$) only above 1000°C. With hydrazine however, complex formation makes it possible to obtain ferrites from the oxalates precursors at low temperatures of $\sim$150°C.

This shows that the exothermic decomposition of hydrazine plays a vital role in the formation of these spinels at low temperatures.

The mixed metal oxalate hydrazinates exhibit autocatalytic combustion behavior and it is of interest to note that the precursors ignite while undergoing suction filtration if allowed to dry.[29] Properties of the ferrites formed are summarized in Table 2.7.

Table 2.7.   Lattice constant, Mössbauer data, and surface area of ferrites.

Ferrite	Lattice constant a (nm)	Mössbauer data		Surface area $(m^2\,g^{-1})$	Particle size[a] (nm)
		Isomer shift $(mm\,s^{-1})$	Internal field (kOe)		
$MgFe_2O_4$	0.842	0.3002	497.7	76	25
$MnFe_2O_4$	0.851	0.3850	524.2	31	61
$CoFe_2O_4$	0.837	0.4790	505.6	47	35
$NiFe_2O_4$	0.835	0.4690	511.0	—	—
$ZnFe_2O_4$	0.844	0.4100	508.4	22	79

[a] From surface area.

Ferrites are widely used as magnetic materials in high-frequency transformer cores, antenna rods, and induction tuners. Ideally, they should have a low value of saturation magnetization and low eddy current loss for high-frequency applications. Losses associated with domain-wall resonance can be minimized when the ferrite is porous. Ferrites also find application as catalysts in dehydrogenation reactions. As a consequence, ferrites with large surface area are needed to function effectively as catalysts. To achieve this, the spinel formation should occur in a short time and the temperature ought to be reduced considerably so that the mixing of the component cations takes place on an atomic scale. Therefore, mixed metal oxalate hydrazinate precursors, $MFe_2(C_2O_4)_3(N_2H_4)_6$, offer a new method for obtaining nano-spinels.

It may be noted that the traditional ceramic approach to synthesize ferrites result in crystalline material but with low surface area, limiting their applications as catalysts and as high-frequency core oxides. The severity of the reaction conditions necessary in the ceramic approach to overcome the slow reaction kinetics between two solids is the cause of this problem. High-temperature firing of the component oxides with repeated regrinding and firing results in crystalline ferrites with low surface area.

2.4.3 *Mixed Metal Oxalate Hydrates: Precursors to Metal Titanates*

Perovskite titanates, $MTiO_3$, where M = Ca, Sr, Ba, or Pb, are important because of their ferroelectric, piezoelectric, and electro-optic properties. Metal titanates ($MTiO_3$) are prepared both by the pyrolysis of the oxalate precursors $MTiO(C_2O_4)_2 \cdot 4H_2O$ (M = Ca, Sr, Ba, or Pb) at 700°C or 1000°C for 1 hour.[30] The stoichiometry of the decomposition of the metal oxalate precursors assuming complete reaction can be written as

$$MTiO(C_2O_4)_2 \cdot 4H_2O(s) \xrightarrow{O_2, 700-1000^0 C} MTiO_3(s) + 4CO_2(g) + 4H_2O(g)$$
(22)

These oxalate precursors can be decomposed at lower temperatures by the addition of calculated amounts of NH_4NO_3 and oxalic acid dihydrazide (ODH) redox mixture. The redox mixture ignites at lower temperatures (350°C) and undergoes combustion.

$$MTiO(C_2O_4)_2 \cdot 4H_2O(s) + C_2H_6N_4O_2(s) + 7NH_4NO_3(s)$$
$$\xrightarrow{350°C} MTiO_3(s) + 6CO_2(g) + 21H_2O(g) + 9N_2 \qquad (23)$$

The addition of NH_4NO_3 not only pyrolyses the oxalate precursor but allows it to undergo combustion autocatalytically, yielding fine particles. This method of enhancing weakly exothermic reactions is called as "Chemical Oven".[31]

The properties of the titanates prepared by pyrolysis and combustion of redox mixtures of oxalate precursors are summarized in Table 2.8.

2.5 MIXED METAL HYDRAZINIUM HYDRAZINE CARBOXYLATES

2.5.1 *Mixed Metal Hydrazinium Hydrazine Carboxylates: Precursors to Nano-Cobaltites and Ferrites*

Mixed metal hydrazinium hydrazine carboxylate precursors, $N_2H_5M_{1/3}Co_{2/3}$ $(N_2H_3COO)_3 \cdot H_2O$ and $N_2H_5M_{1/3}Fe_{2/3}(N_2H_3COO)_3 \cdot H_2O$ (where M = Mg, Mn, Fe, Co, Ni, Zn, and Cd), are prepared by the reaction of an

Table 2.8. Composition of redox mixtures and particulate and dielectric properties of $MTiO_3$ (M = Ca, Sr, Ba, or Pb).

Oxalate precursors		Tap density ($g\,cm^{-3}$)	Powder density (% theoretical)[a]	Surface area ($m^2\,g^{-1}$)	Particle size[b] (μm)	Room temp. dielectric constant at 10 kHz (1100°C sintered pellet)
$CaTiO_3^*$	A	0.12	3.20 (78%)	1.70	1.10	101
	B	0.06	2.58 (63%)	18.80	0.12	128
$SrTiO_3^\#$	A	0.08	4.10 (80%)	16.50	0.08	363
	B	0.03	2.51 (51%)	28.50	0.08	428
$BaTiO_3^\#$	A	0.09	5.08 (84%)	5.50	0.21	1620
	B	0.03	4.40 (73%)	42.00	0.03	1965
$PbTiO_3^*$	A	0.14	6.69 (87%)	2.50	0.38	196
	B	0.08	5.25 (70%)	30.00	0.04	235

A. $MTiO_3$ obtained by heating 5 g oxalate precursor at 700°C* and 1000°C# for 1 h.
B. $MTiO_3$ obtained from combustion of 5 g oxalate precursors with:
7.955 g NH_4NO_3 + 1.683 g ODH for M = Ca,
7.008 g NH_4NO_3 + 1.477 g ODH for M = Sr,
6.160 g NH_4NO_3 + 1.322 g ODH for M = Ba,
5.393 g NH_4NO_3 + 1.137 g ODH for M = Pb.
[a]Theoretical density: $CaTiO_3$ 4.018 $g\,cm^{-3}$, $SrTiO_3$ 5.11 $g\,cm^{-3}$, $BaTiO_3$ 6.017 $g\,cm^{-3}$, and $PbTiO_3$ 7.52 $g\,cm^{-3}$.
[b]From surface area.

aqueous solution of metal sulfates containing $M^{2+}:Co/Fe^{2+}$ ions in the mole ratio of 1:2 with $N_2H_3COON_2H_5$ or N_2H_3COOH in $N_2H_4 \cdot H_2O$.[32–34] Crystalline solids separate from the solution in a couple of days. The crystals are washed with alcohol and dried over P_2O_5 in a vacuum desiccator.

These complexes are characterized by their chemical analysis and XRDs (Fig. 2.17). Powder X-ray diffraction patterns of all the precursors are identical, indicating solid solution formation. The unit cell parameters and the volume calculated from the powder XRD are shown in Table 2.9. Although $N_2H_5M_{1/3}Fe_{2/3}(N_2H_3COO)_3 \cdot H_2O$, where M = Mg, and Mn are not known, solid-solution precursors containing Mg^{2+} and Mn^{2+} are formed due to their comparable ionic radii with that of Fe^{2+} ion (i.e., Mg^{2+} = 0.065 nm, Mn^{2+} = 0.080 nm, and Fe^{2+} = 0.075 nm).

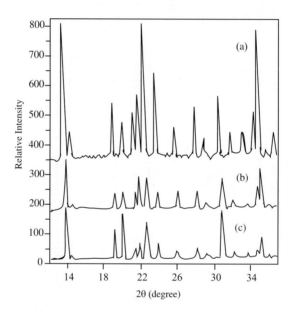

Fig. 2.17. XRD patterns of: (a) $N_2H_5Zn(N_2H_3COO)_3 \cdot H_2O$, (b) N_2H_5Co $(N_2H_3COO)_3 \cdot H_2O$, and (c) $N_2H_5Zn_{1/3}Co_{2/3}(N_2H_3COO)_3 \cdot H_2O$.

All the precursors decompose exothermically in more than one step at temperatures between 120°C and 250°C to yield corresponding metal cobaltites and ferrites. A typical TG/DTA of the ferrite precursor is given in Fig. 2.18. The ferrite precursors decompose in a single step and at much lower temperatures between 75°C and 200°C as compared to the cobaltite precursors.

Table 2.9. Unit cell dimensions for monoclinic hydrazinium metal hydrazine carboxylate hydrates and their solid solutions.

Complex	a (nm)	b (nm)	c (nm)	β°	V (nm³)
$N_2H_5Ni(N_2H_3COO)_3 \cdot H_2O^a$	1.2129	1.0858	1.0255	121.10	1.1564
$N_2H_5Fe(N_2H_3COO)_3 \cdot H_2O$	1.2106	1.0944	1.0286	120.68	1.1720
$N_2H_5Co(N_2H_3COO)_3 \cdot H_2O$	1.2106	1.0873	1.0252	121.10	1.1550
$N_2H_5Mg_{1/3}Fe_{2/3}(N_2H_3COO)_3 \cdot H_2O$	1.2151	1.1007	1.0282	120.68	1.1826
$N_2H_5Mn_{1/3}Fe_{2/3}(N_2H_3COO)_3 \cdot H_2O$	1.2165	1.1029	1.0272	121.69	1.1726
$N_2H_5Co_{1/3}Fe_{2/3}(N_2H_3COO)_3 \cdot H_2O$	1.2207	1.1016	1.0268	121.09	1.1824
$N_2H_5Ni_{1/3}Fe_{2/3}(N_2H_3COO)_3 \cdot H_2O$	1.2111	1.1016	1.0268	121.53	1.1677
$N_2H_5Ni_{1/3}Fe_{2/3}(N_2H_3COO)_3 \cdot H_2O^a$	1.1850	1.0960	1.0260	118.21	1.1742
$N_2H_5Zn_{1/3}Fe_{2/3}(N_2H_3COO)_3 \cdot H_2O$	1.2130	1.1006	1.0279	120.73	1.1795

[a] Single crystal data.[33]

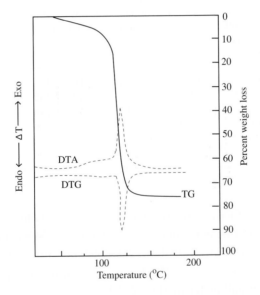

Fig. 2.18. Simultaneous TG–DTA–DTG of $N_2H_5Co_{1/3}Fe_{2/3}(N_2H_3COO)_3 \cdot H_2O$.

The formation of the spinel cobaltites and ferrites was confirmed by powder XRD patterns, which show the characteristic X-ray line broadening due to the fine particle nature of the products. The lattice constants (a values) and crystallite sizes calculated from the XRD patterns for cobaltites and ferrites are summarized in Tables 2.10 and 2.11, respectively.

Table 2.10.   Particulate properties of nano-cobaltites.

Metal	Lattice constant a (nm)	Surface area ($m^2 g^{-1}$)	Crystallite size[a] (nm)	Conductivity ($\Omega^{-1} cm^{-1}$)
Mg	0.8130	47	10	10^{-4}
Mn	0.8113	24	24	10^{-5}
Fe	0.8222	116	7	10^{-3}
Co	0.8108	18	28	10^{-5}
Ni	0.8118	12	10	10^{-2}
Zn	0.8172	65	14	10^{-6}

[a] From XRD.

Table 2.11.   Magnetic and particulate properties of nano-ferrites.

Ferrite	Mössbauer data		Lattice constant a (nm)	Particle density ($\times 10^3$) ($kg\,m^{-3}$)	Surface area ($m^2 g^{-1}$)	Crystallite size from XRD (nm)	Particle size[a] (nm)
	Isomer shift stainless steel ($mm\,s^{-1}$)	Internal field (kOe)					
$MgFe_2O_4$	0.385	507.0	0.8388	4.118	114	13	13
$MnFe_2O_4$	0.337	522.0	0.8321	3.269	140	6	14
$CoFe_2O_4$	0.467	548.0	0.8418	3.649	116	10	15
$NiFe_2O_4$	0.457	524.0	0.8359	3.574	26	22	68
$ZnFe_2O_4$	0.409	—	0.8477	3.347	108	9	17
$CdFe_2O_4$	—	—	0.8698	3.458	93	—	20

[a] From surface area.

Surface areas of ferrites are extremely high and crystallite sizes determined by Scherrer method are in the range of 10–20 nm. The Mössbauer spectrum of $MgFe_2O_4$ at room temperature shows a doublet indicating superparamagnetic behavior, which on sintering at 250°C and 500°C gives the expected six-finger pattern of the crystalline ferrite (Fig. 2.19).

Technologically, nanosize ferrites have been of interest due to their applications in the preparation of high-density ferrites at low temperature, e.g., $ZnFe_2O_4$ when sintered at 1000°C for 24 h achieved 99% theoretical density (Fig. 2.20). This temperature is quite low compared to the normally employed temperature of 1200–1400°C.[33]

Nanosize ferrites are also used as suspension materials in ferromagnetic liquids, and as catalysts. The catalytic activity of both nano-cobaltites and

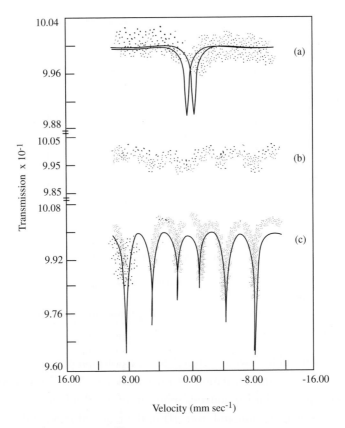

Fig. 2.19. Mössbauer spectra of ^{57}Fe in $MgFe_2O_4$ at (a) 250°C, (b) sintered at 500°C, 24 h, and (c) sintered at 900°C, 24 h. The solid line is the least square fit.

ferrites has been investigated for H_2O_2 decomposition. Cobaltites have been found to be better catalysts than ferrites.[35]

2.5.2 *Mixed Metal Hydrazinium Hydrazine Carboxylates: Precursors to Mixed Ferrites*

Among the nano-ferrites, mixed metal ferrites are also important. Of these, nickel zinc ferrites ($Ni_{1-x}Zn_xFe_2O_4$, $0 < x < 1$) have been widely used in high-frequency applications; high-density Mn–Zn ferrites ($Mn_{1-x}Zn_xFe_2O_4$, $0 < x < 1$) are used in read write heads, high-speed digital tapes, broad band transformer core, etc. These mixed ferrites can be prepared by the above

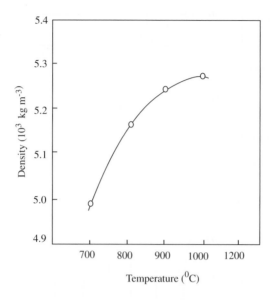

Fig. 2.20.   Plot of density versus sintering temperature for $ZnFe_2O_4$.

hydrazinium metal hydrazine carboxylate solid-solution precursor technique. The precursors for the formation of mixed Ni–Zn, Mn–Zn and Mg–Mn ferrites were prepared using aqueous solutions of metal sulfate of nickel, zinc, manganese or magnesium, and iron (II) in the required ratio, and a solution of N_2H_3COOH in $N_2H_4 \cdot H_2O$ and investigated.[36–39]

Thermal decomposition studies of these precursors show that they decompose exothermally in a single step in the temperature range of 120–250°C. The precursors of Ni–Zn ferrites decompose at 250°C by heating in an oxygen atmosphere for 1 h, while that of Mn–Zn decompose at ~170°C. The properties of Ni–Zn ferrite prepared by the combustion of single source precursors are summarized below (Table 2.12).

Permeability measurements have been made on all the ferrites sintered at 900°C, 1000°C, and 1100°C in the frequency range of 1 kHz to 10 MHz. Magnetic properties of the Ni–Zn ferrites are summarized in Table 2.13.

It is clear from Table 2.13 that the density of the ferrite does not change much with the firing temperature but the value of μ increases with the increase in sintering temperature. This could be attributed to the variation in grain size of the ferrite.[37]

Table 2.12. Properties of Ni–Zn ferrite.

Precursors	Oxide	Surface area (m^2g^{-1})	Particle size* (nm)
$(N_2H_5)_3Ni_xZn_{1-x}$ $Fe_2(N_2H_3COO)_3 \cdot H_2O$			
$x = 0.2$	$Ni_{0.2}Zn_{0.8}Fe_2O_4$	91	21
$x = 0.4$	$Ni_{0.4}Zn_{0.6}Fe_2O_4$	86	22
$x = 0.5$	$Ni_{0.5}Zn_{0.5}Fe_2O_4$	90	21
$x = 0.6$	$Ni_{0.6}Zn_{0.4}Fe_2O_4$	85	22
$x = 0.8$	$Ni_{0.8}Zn_{0.2}Fe_2O_4$	91	21

*From surface area.

Table 2.13. Permeability (μ) of $Ni_{0.75}Zn_{0.25}Fe_2O_4$ as a function of temperature and frequency.

Sintering temperature (°C)	Theoretical density (%)	Permeability (μ) as a function of frequency				Grain size (μm)
		1 kHz	10 kHz	1 MHz	10 MHz	
900	98.4	23.72	48.85	49.31	49.77	<1
1000	99.5	50.89	70.43	69.07	73.16	2
1100	99.5	138.84	140.67	129.6	121.71	4

Using this method, one can obtain high-density ($\sim$99%) ferrites at relatively low sintering temperatures (1100°C) and thus can avoid hot-pressing techniques because of the nanosize of ferrite particles ($\sim$60 nm).

2.5.3 *Mixed Metal Hydrazinium Hydrazine Carboxylates: Precursors to Manganites*

Fine particle spinel manganites have been prepared by the thermal decomposition of the precursors $N_2H_5M_{1/3}Mn_{2/3}(N_2H_3COO)_3 \cdot H_2O$ (M = Co and Ni) and $M_{1/3}Mn_{2/3}(N_2H_3COO)_2 \cdot 2H_2O$ (M = Mg and Zn). Once ignited between 250–375°C, these precursors undergo self-propagating gas producing exothermic reactions to yield corresponding manganites in less than 5 min.[40] Formation of a single phase in the decomposed product is confirmed by powder XRD patterns. The manganites are of submicrometer size and have surface area in the range of 20–76 $m^2 g^{-1}$. Table 2.14 summarizes the properties of manganites.

Table 2.14.   Properties of manganites.

Precursor	DTA peak temperature (°C)	Product	Surface area (m^2g^{-1})	Particle size[a] (μm)
$Mg_{1/3}Mn_{2/3}(N_2H_3COO)_2 \cdot 2H_2O$	225 (exo)	$MgMn_2O_4$	28	4.85
$Zn_{1/3}Mn_{2/3}(N_2H_3COO)_2 \cdot 2H_2O$	240 (exo)	$ZnMn_2O_4$	61	0.69
$N_2H_5Co_{1/3}Mn_{2/3}(N_2H_3COO)_3 \cdot H_2O$	205 (exo)	$CoMn_2O_4$	76	—
$N_2H_5Ni_{1/3}Mn_{2/3}(N_2H_3COO)_3 \cdot H_2O$	210 (exo)	$NiMn_2O_4$	20	1.90

[a] From surface area.

Several other mixed metal oxides like chromites (MCr_2O_4), ferrites (MFe_2O_4), and titanates ($Pb(Zr/Ti)O_3$) have been prepared using combustible single source precursors.[41] The precursors employed are (N_2H_5) $MCr_2(C_2O_4)_5 \cdot 3N_2H_4$,[15] $MFe_2(SO_3)_3 \cdot 6N_2H_4 \cdot 2H_2O$,[42] and $PbTiO(OH)_2$ ($N_2H_3COO)_2 \cdot H_2O$, $PbZrO(OH)_2(N_2H_3COO)_2 \cdot H_2O$[43] respectively.

Recently, single-source precursor route to $LiMO_2$ (M = Ni, Co) battery materials has been prepared by pyrolysis of $[Li(H_2O)M(N_2H_3CO_2)_3] \cdot 0.5H_2O$ precursors at ~700°C in air.[44]

2.6 CONCLUDING REMARKS

Hydrazine has been successfully employed to prepare variety of metal complexes. These complexes are used as single-source solid combustible precursors to produce nanosize oxides directly. Technologically useful oxides like cobaltites, ferrites, manganites, etc. have been prepared by low-temperature combustion of hydrazinium carboxylate based precursors.

References

1. Rao CNR, *Chemical Approaches to the Synthesis of Inorganic Materials*, Wiley Eastern Ltd., New Delhi, pp. 18–27, 1994.
2. Whipple E, Wold A, Preparation of stoichiometric chromite, *J Inorg Nucl Chem* **24**: 23–27, 1962.

3. Vidyasagar K, Gopalkrishnan J, Rao CNR, Synthesis of complex metal oxides using hydroxide, cyanide and nitrate solid solution precursors, *J Solid State Chem* **58**: 29–37, 1985.

4. Longo JM, Horowitz HS, Solid state synthesis of complex oxides, in Honig JM, Rao CNR (eds.), *Preparation and Characterization of Materials*, Academic Press, New York, pp. 29–46, 1981.

5. Sekar MA, Dhanaraj G, Bhat HL, Patil KC, Synthesis of fine particle titanates by the pyrolysis of oxalate precursors, *J Mater Sci (Mater Electron)* **3**: 237–239, 1992.

6. Laurent Ch, Rousset A, Verelst M, Kannan KR, Raju AR, Rao CNR, Reduction behaviour of Fe^{3+}/Al_2O_3 obtained from the mixed oxalate precursor and the formation of the Fe^0-Al_2O_3 metal–ceramic composite, *J Mater Chem* **3**: 513–518, 1993.

7. Horowitz HS, Mchain SJ, Sleight QW, Druliner JD, Gai PL, Van Kavaelaur MJ, Wagner JL, Biggs BD, Poon SJ, Submicrometer superconducting particles made by a low temperature synthetic route, *Science* **243**: 66–69, 1989.

8. Roesky HW, Mockel K, *Chemical Curiosities*, Mitchell TN, Russey WE (English translators) VCH Publishers Inc., New York, pp. 78–79, 1996.

9. Audreith LF, Ogg BA, *The Chemistry of Hydrazine*, Wiley, New York, 1951.

10. Schmidt EW, *Hydrazine and its Derivatives*, Wiley, New York, 1984.

11. Patil KC, Budkuley JS, Pai Vernekar VR Thermoanalytical studies of hydrazido carbonic acid derivatives, *J Inorg Nucl Chem* **41**: 953–955, 1979.

12. Patil KC, Soundararajan R, Goldberg EP, Reaction of hydrazinium hydrazido-carboxylate salts of Mg, Mn, Fe, Co, Ni and Zn, *Synth React Inorg Met-Org Chem* **13**: 29–43, 1983.

13. Ravindranath P, Patil KC, Preparation, characterization and thermal analysis of metal hydrazine carboxylate derivatives, *Proc Ind Acad Sci (Chem Sci)* **95**: 345–356, 1985.

14. Patil KC, Metal hydrazine complexes as precursors to oxide materials, *Proc Ind Acad Sci (Chem Sci)* **96**: 459–464, 1986.

15. Patil KC, Contributions to the chemistry of hydrazine derivatives, *Symposium on Hydrazine Centenary, Indian Institute of Science, Bangalore*, pp. 48–53, 1987.

16. Patil KC, Nesmani C, Pai Vernekar VR, Synthesis and characterization of metal hydrazine nitrate, azide and perchlorate complexes, *Synth React Inorg Met-Org Chem* **12**: 383–395, 1982.

17. Kishore K, Patil KC, Gajapathi D, Mechanistic studies on self-deflagrating solids from temperature profile analysis, *Propellant Explosive Pyrotechnics* **10**: 187–191, 1985.

18. Ravindranathan P, Patil KC, Thermal reactivity of metal formate hydrazinates, *Thermochimica Acta* **71**: 53–57, 1983.

19. Mahesh GV, Patil KC, Thermal reactivity of metal acetate hydrazinates, *Thermochimica Acta* **99**: 153–158, 1986.
20. Patil KC, Gajapathy D, Kishore K, Thermal reactivity of metal oxalate hydrazinates, *Thermochimica Acta* **52**: 113–120, 1982.
21. Mahesh GV, Ravindranathan P, Patil KC, Preparation, characterization and thermal analysis of rare earth and uranyl hydrazinecarboxylate derivatives, *Proc Ind Acad Sci (Chem Sci)* **97**: 117–123, 1986.
22. Sekar MMA, Patil KC, Hydrazine carboxylate precursors to fine particle titania, zirconia and zirconium titanate, *Mat Res Bull* **28**: 485–492, 1983.
23. Ravindranathan P, Patil KC, A one step process for the preparation of γ-Fe$_2$O$_3$, *J Mater Sci Lett* **5**: 221–222, 1986.
24. Suresh K, Mahesh GV, Patil KC, Preparation of cobalt substituted gamma-Fe$_2$O$_3$, in Srivastava CM, Patni MJ (eds.), *Advances in Ferrites*, Oxford & IBH Pub. Co. Pvt. Ltd, New Delhi, pp. 893–897, 1989.
25. Aruna ST, Ghosh S, Patil KC, Combustion synthesis and properties of Ce$_{1-x}$Pr$_x$O$_{2-\delta}$ red ceramic pigments, *Int J Inorganic Maters* **3**: 387–392, 2001.
26. Ekambaram S, Patil KC, Maaza M, Synthesis of lamp phosphors: Facile combustion approach, *J Alloys Comp* **393**: 81–92, 2005.
27. Mahesh GV, Patil KC, Low temperature preparation of nickel and zinc cobaltites, *Reactivity Solids* **4**: 117–123, 1987.
28. Patil KC, Gajapathy D, Pai Vernekar VR, Low temperature cobaltite, formation using mixed metal oxalate hydrazinate precursor, *J Mater Sci Lett* **2**: 272–274, 1983.
29. Gajapathy D, Patil KC, Mixed metal oxalate hydrazinates as compound precursors to spinel ferrites, *Mater Chem Phys* **9**: 423–438, 1983.
30. Sekar MMA, Dhanraj G, Bhat HL, Patil KC, Synthesis of fine particle titanates by the pyrolysis of oxalate precursors, *J Mater Sci Mater Electron* **3**: 237–239, 1992.
31. Munir ZA, Synthesis of high temperature materials by self-propagating combustion methods, *Ceram Bull* **67**: 342–349, 1988.
32. Ravindranathan P, Mahesh GV, Patil KC, Low-temperature preparation of fine particle cobaltites, *J Solid State Chem* **66**: 20–25, 1987.
33. Ravindranathan P, Patil KC, A low temperature path to ultrafine ferrites, *Am Ceram Soc Bull* **66**: 688–692, 1987.
34. Arul Dhas N, Muthuraman M, Ekambaram S, Patil KC, Synthesis and properties of fine particle cadmium ferrite (CdFe$_2$O$_4$), *Intl J SHS* **3**: 39–50, 1994.
35. Mimani Tanu, Ravindranathan P, Patil KC, Catalytic decomposition of hydrogen peroxide on fine particle ferrites and cobaltites, *Proc Indian Acad Sci (Chem Sci)* **99**: 209–215, 1987.

36. Ravindranathan P, Patil KC, Novel solid solution precursor method for the preparation of ultrafine Ni-Zn ferrites, *J Mater Sci* **22**: 3261–3264, 1987.

37. Srinivasan TT, Ravindranathan P, Cross LE, Roy R, Newnham RE, Sankar SG, Patil KC, Studies on high-density nickel zinc ferrites and its magnetic properties using novel hydrazine precursors, *J Appl Phy* **63**: 3789–3791, 1988.

38. Suresh K, Patil KC, Preparation of high density Mn–Zn ferrites, in Srivastava CM, Patni MJ (eds.), *Advances in Ferrites*, Oxford & IBH Pub. Co. Pvt. Ltd., New Delhi, pp. 103–107, 1989.

39. Sundar Manoharan S, Patil KC, Preparation and properties of fine particle Mg–Mn ferrites, in Srivastava CM, Patni MJ (eds.), *Advances in Ferrites*, Oxford & IBH Pub. Co. Pvt. Ltd., New Delhi, pp. 43–47, 1989.

40. Arul Dhas N, Patil KC, Combustion synthesis and properties of fine particle manganites, *J Solid State Chem* **102**: 440–445, 1993.

41. Patil KC, Sekar MMA, Synthesis, structure and reactivity of metal hydrazine carboxylates: Combustible precursors to fine particle oxide materials, *Int J Self-Propagating High-Temp Synth* **3**: 181–196, 1994.

42. Budkuley JS, Patil KC, Synthesis and thermoanalytical properties of mixed metal sulfite hydrazinate hydrates I, *Synth React Inorg Met Org Chem* **19**: 909–922, 1989.

43. Sekar MMA, Patil KC, Combustion synthesis of lead-based dielectrics: A comparative study of redox compounds and mixtures, *Int J Self-Propagating High-Temp Synth* **3**: 27–38, 1989.

44. Tey SL, Reddy MV, Subba Rao GV, Chowdary BVR, Ding J Yi J, Vittal JJ, Synthesis, structure, and magnetic properties of $[Li(H_2O)M(N_2H_3CO_2)_3] \cdot 0.5H_2O$ (M = Co, Ni) as single precursors to $LiMO_2$ battery materials, *Chem Mater* **18**: 1587–1594, 2006.

Chapter 3

Solution Combustion Synthesis of Oxide Materials

3.1 INTRODUCTION

Combustion synthesis or fire synthesis is also known as self-propagating high-temperature synthesis (SHS). To generate fire, an oxidizer, a fuel, and the right temperature are needed. All these three elements make up a fire triangle. Fire can be described as an uncontrolled combustion, which produces heat, light, and ash (Fig. 3.1). The process makes use of highly exothermic redox chemical reactions between an oxidizer and a fuel.

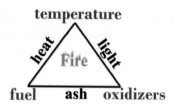

Fig. 3.1.   The fire triangle.

All self-propagating high-temperature synthesis reactions are redox reactions; however, all redox reactions need not be SHS reactions. For the reaction to be self-propagating the heat evolved should be more than the heat required for initiating the combustion. A redox reaction involves simultaneous oxidation and reduction processes. The classical definition of oxidation is addition of oxygen or any other electronegative element (nonmetal), while reduction is addition of hydrogen or any other electropositive element (metal). The term combustion covers flaming (gas-phase), smouldering (solid-gas) as well as explosive reactions. It can be linear combustion or volume combustion. In linear combustion the burning surface recedes from top to bottom in layers, whereas in volume combustion, the entire reaction mixture ignites to burn with a flame.

The SHS process pioneered by Merzhanov has been used to prepare a large number of technologically useful oxide materials (refractories, magnetic, semi-conductors, dielectric, catalysts, sensors, phosphors, etc.) and nonoxide materials (carbides, nitrides, borides, silicides) by the solid-state reaction between the corresponding metals and nonmetals.[1] The process requires high-purity fine precursors, which ignite at temperatures $>1000°C$. The process is highly exothermic ($T_{ad} \sim 4000°C$) and self-propagating resulting in coarse products. This process has been successfully used in Russia to prepare hundreds of technologically useful materials,[2] including the preparation of oxide materials in recent times. However, being a solid-state reaction it often does not produce homogeneous products and results in coarse powders. Here, the solution combustion approach is presented to prepare oxide materials of desired composition, structure, and properties.

3.2 SOLUTION COMBUSTION SYNTHESIS (SCS)

SCS of oxide materials was unexpectedly discovered during the reaction between aluminum nitrate and urea. A mixture of $Al(NO_3)_3 \cdot 9H_2O$ and urea solution, when rapidly heated around $500°C$ in a muffle furnace, foamed and ignited to burn with an incandescent flame yielding voluminous white product which was identified as α-Al_2O_3.[3]

To understand the highly exothermic nature of this reaction, concepts used in propellant chemistry were employed.[4] A solid propellant contains an oxidizer like ammonium perchlorate and a fuel like carboxy terminated polybutadiene together with aluminum powder and some additives. The specific impulse (I_{sp}) of a propellant, which is a measure of energy released during combustion is given by the ratio of thrust produced per pound of the propellant. It is expressed as

$$I_{sp} = k\sqrt{\frac{T_c}{\text{Molecular Wt. of gaseous products}}}. \tag{1}$$

The highest heat T_c (chamber temperature in the rocket motor) is produced when the equivalence ratio (Φ_e = oxidizer/fuel ratio) is unity.

The equivalence ratio of an oxidizer and fuel mixture is expressed in terms of the elemental stoichiometric coefficient.

$$\Phi_e = \frac{\sum (\text{Coefficient of oxidizing elements in specific formula}) \times (\text{Valency})}{(-1) \sum (\text{Coefficient of reducing elements in specific formula}) \times (\text{Valency})}. \qquad (2)$$

A mixture is said to be stoichiometric when $\Phi_e = 1$, fuel lean when $\Phi_e > 1$, and fuel rich when $\Phi_e < 1$. Stoichiometric mixtures produce maximum energy.

The oxidizer/fuel molar ratio (O/F) required for a stoichiometric mixture ($\Phi_e = 1$) is determined by summing the total oxidizing and reducing valencies in the oxidizer compounds and dividing it by the sum of the total oxidizing and reducing valencies in the fuel compounds. In this type of calculation oxygen is the only oxidizing element; carbon, hydrogen, and metal cations are reducing elements and nitrogen is neutral. Oxidizing elements have positive valencies and reducing elements have negative valencies.[5]

In solution combustion calculations, the valency of the oxidizing elements was modified and considered as negative, and the reducing elements as positive, similar to the oxidation number concept familiar to chemists. Accordingly, the elemental valency of C, Al, and H is +4, +3, and +1, respectively, and oxidizing valency of oxygen is taken as −2. The valency of nitrogen is considered to be zero. Based on this concept, the oxidizing valency of aluminum nitrate and the reducing valency of urea are

$$Al(NO_3)_3 = -15; \quad [Al = +3, 3N = 0, 9O = (9 \times -2) = -18],$$
$$CH_4N_2O = +6; \quad [C = +4, 4H = (4 \times +1) = +4, 2N = 0,$$
$$O = (1 \times -2) = -2].$$

Accordingly, for the complete combustion of aluminum nitrate : urea mixture, the molar ratio becomes 15/6 = 2.5. The stoichiometric equation for this

reaction can be written as

$$2Al(NO_3)_3(aq) + 5\underset{\text{Urea}}{CH_4N_2O(aq)} \xrightarrow{500°C} \alpha\text{-}Al_2O_3(s) + 5CO_2(g)$$
$$+ 8N_2(g) + 10H_2O(g) \qquad\qquad (3)$$

When the reaction of aluminum nitrate and urea is carried out in the molar ratio of 1 : 2.5 the energy released is maximum and the combustion is complete with no carbon residue. This type of stoichiometric balance of a redox mixture for a combustion reaction is fundamental to the synthesis of an oxide material by the solution combustion method.

3.2.1 *Synthesis of Alumina*

To synthesize alumina as per the above molar ratio, $Al(NO_3)_3 \cdot 9H_2O$ (20.0 g) and urea (8.0 g) are dissolved in 15 ml of distilled water. The mixture is taken in a 300 cm³ Pyrex dish and introduced into a muffle furnace (38 × 17 × 9 cm³) maintained at 500°C. The mixture boils, foams, ignites, and burns with an incandescent flame of temperature ~1500°C as measured by disappearing filament optical pyrometer. The product of combustion is $\alpha\text{-}Al_2O_3$, which occupies the entire volume of the container (weight 2.7 g, yield 100%) (Fig. 3.2). The entire combustion process is completed in about 3 min.

Fig. 3.2. $\alpha\text{-}Al_2O_3$ prepared by solution combustion method.

This form of combustion synthesis offers some unique features. The combustion process is completed within a short span of just ∼150–180 s with flame temperature reaching as high as ∼1500°C as measured by the optical pyrometer. Interestingly, there is no damage to the Pyrex container in which the reaction is carried out; probably due to the thermal insulating nature of alumina and short duration of combustion time. The combustion residue weighing 2.72 g occupies the entire volume of the container (300 cm^3) resulting in a very light powder of density ∼0.009 g cm^{-3}. Evolution of large quantities of gases like carbon dioxide, water, and nitrogen results in the formation of fine particle alumina. The average particle size and surface area of alumina are 4 μm and 8 m^2 g^{-1}, respectively.

The observed high flame temperature (1500°C) of the reaction is due to the exothermicity of the reaction between the gaseous decomposition products of $Al(NO_3)_3$ as NO_x and urea as HNCO, NH_3, etc. These decomposition products of aluminum nitrate and urea were identified by TG–DTA and EGA–MS studies. The TG–DTA data show melting of urea at 100–150°C and its exothermic decomposition at 400°C and 500°C. EGA and mass spectrometry data of urea show m/e at 17, 43, and 44 corresponding to the formation of NH_3, HNCO, and CO_2 species, respectively. These gaseous decomposition products of urea are known to be hypergolic with nitrogen oxides, i.e., once they attain a critical density and the required temperature, they burn with a flame (temperature 1500 ± 100°C) even at ambient pressures. The presence of HNCO as one of the decomposition products appears to give rise to flame with high temperatures. Reactions of nitrogen oxides with ammonia are known to give flame temperature of about 2000°C.[6]

3.2.2 *Mechanism of Aluminum Nitrate — Urea Combustion Reaction*

Based on the knowledge of the thermal and combustion behavior of aluminum nitrate, urea, and their mixtures, a probable mechanism for SCS is proposed.

When a mixture containing $Al(NO_3)_3 \cdot 9H_2O$ and urea with required stoichiometry of 1 : 2.5 is heated rapidly at 500°C, it undergoes melting and dehydration in the first 2 min. Later it decomposes with frothing as a result of the formation of $Al(OH)(NO_3)_2$ gel along with other products like urea nitrate, biuret, HNCO, and NH_3. This mixture then foams due to the generation of gaseous decomposition products as intermediates, leading to enormous swelling. The gaseous decomposition products are a mixture of nitrogen oxides, NH_3, and HNCO. These gases are known to be hypergolic in contact with each other. The foam could be made up of polymers like cyanuric acid, polymeric nitrate, etc., which are combustible. In the third minute, the foam breaks out with a flame because of the accumulation of the hypergolic mixture of gases. With an *in situ* temperature build up of 1350°C the whole foam further swells and burns to incandescence. At such high *in situ* temperature the foam decomposes to yield 100% α-Al_2O_3 (the high-temperature form). This proposed mechanism is illustrated below (Fig. 3.3).

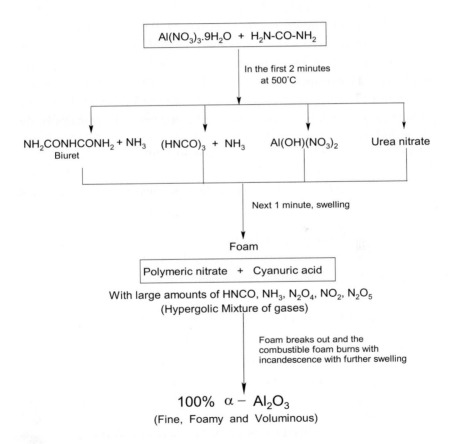

Fig. 3.3.   Mechanism of solution combustion synthesis of α-Al_2O_3.

However, in cases where the gases responsible for combustion are allowed to escape or when their densities are less or the required temperature is not met, combustion does not occur with a flame. The foam only sustains combustion but itself does not initiate ignition. Also, if the mixture is heated at a slow rate the flame does not appear, as the time required to reach the ignition temperature is longer and all the gases responsible for combustion escape the foam.

3.2.3 Thermodynamic Calculation

This discussion has been further supported by theoretical calculations of the combustion reaction between aluminum nitrate and urea based on the heat of formation of reactants and products.

In solution combustion reactions, extremely high temperatures of over 1550°C can be achieved within a very short duration. Therefore, it is reasonable to assume that a thermally isolated system exists because there is very little time for the heat to disperse to its surroundings. Consequently, the maximum temperature to which the product is raised is assumed to be adiabatic temperature (T_{ad}). The heat liberated during the reaction is the enthalpy of the system and is a state function. It is expressed as

$$\Delta H^0 = \Delta H_f^0 = \int_{298}^{T_{ad}} \Delta C_p(\text{product}) \, dT. \tag{4}$$

The heat liberated during the combustion synthesis of alumina from aluminum nitrate–urea mixtures can be calculated using the ΔH_f^0 values of the reactants and products (heat of formation in kcal mol^{-1}) at STP.

Reactants: $Al(NO_3)_3 \cdot 9H_2O = -897.96$; Urea $= -79.71$,
Products: α-$Al_2O_3 = -400.4$; $CO_2 = -94.051$; $H_2O(\text{water}) = -57.796$.[7]

The C_p value for α-$Al_2O_3 = 18.89$ cal deg^{-1} mol^{-1}

$$2Al(NO_3)_3 \cdot 9H_2O + 5CH_4N_2O \xrightarrow{773\,K} \alpha\text{-}Al_2O_3(s)$$
$$+5CO_2(g) + 8N_2(g) + 28H_2O(g) \tag{5}$$

For a thermally isolated (adiabatic) system $\Delta H_f^0 = 0$

$$\Delta H_f^0 (\text{reactants})$$
$$= \Delta H_f^0 (\text{Al(NO}_3)_3 \cdot 9H_2O) + \Delta H_f^0 (\text{Urea})$$
$$= 2 \times -897.96 + 5 \times -79.71 = -2194.47 \, \text{kcal mol}^{-1}, \quad (6)$$

$$\Delta H_f^0 (\text{products})$$
$$= \Delta H_f^0 (\alpha\text{-Al}_2O_3) + \Delta H_f^0 (CO_2) + \Delta H_f^0 (N_2) + \Delta H_f^0 (H_2O)$$
$$= -400.4 + 5 \times -94.05 + 9 \times 0 + 28 \times -57.796$$
$$= -2488.943 \, \text{kcal mol}^{-1}, \quad (7)$$

$$\Delta H_f^0 (\text{reaction}) = \Delta H_f^0 (\text{products}) - \Delta H_f^0 (\text{reactants})$$
$$= -2488.93 - (-2194.47) = -294.47 \, \text{kcal mol}^{-1}. \quad (8)$$

Using ΔH_f^0 (reaction) one can calculate the flame temperature:

$$-(-294.47) = \Delta H^0 = \Delta H_f^0 = \int_{298}^{T_{ad}} \Delta C_{p(\text{product})} \, dT. \quad (9)$$

Solving the above equation, a value of 1700 K was obtained for T_{ad}, which agrees very well with the observed $T_{flame}(1773 \pm 100 \, \text{K})$.

3.3 ROLE OF FUELS

Urea is documented as an ideal fuel for the combustion synthesis of high-temperature oxides like alumina and alkaline earth aluminates. However, a need to employ alternate fuels to prepare oxides which are unstable above 1000°C such as transition metal aluminates was necessary. In this regard, hydrazine-based fuels like carbohydrazide (CH), oxalyl dihydrazide (ODH), and malonic dihydrazide (MDH), which have low ignition temperature and are combustible due to the presence of N–N bond that decomposes exothermically to N_2 (N≡N) were found to be suitable. The list of fuels investigated and used has been summarized in Table 3.1.

Table 3.1. Commonly used fuels in solution combustion synthesis.

Sl. No.	Fuel	Formula	Reducing valency	Reactants for preparation	Structure
1.	Urea (U)	CH_4N_2O	+6	Commercially available	
2.	Glycine (G)	$C_2H_5NO_2$	+9	Commercially available	
3.	Hexamethylene tetramine (HMT)	$C_6H_{12}N_4$	+36	$4NH_3 + 6HCHO$ Commercially available	
4.	Carbohydrazide (CH)	CH_6N_4O	+8	$(C_2H_5)_2CO + 2N_2H_4H_2O$ (Ref. 8)	
5.	Oxalic acid dihydrazide (ODH)	$C_2H_6N_4O_2$	+10	$(COOC_2H_5)_2 +$ $2N_2H_4H_2O$ (Ref. 9)	

(*Continued*)

Table 3.1. (*Continued*)

Sl. No.	Fuel	Formula	Reducing valency	Reactants for preparation	Structure
6.	Malonic acid dihydrazide (MDH)	$C_3H_8N_4O_2$	+16	$CH_2(COOC_2H_5)_2 +$ $2N_2H_4H_2O$ (Ref. 10)	
7.	Maleic hydrazide (MH)	$C_4H_4N_2O_2$	+16	$+ N_2H_6Cl_2$ (Ref. 11)	
8.	Diformyl hydrazide (DFH)	$C_2H_4N_2O_2$	+8	$2HCOOH + N_2H_4H_2O$ (Ref. 12)	
9.	Tetraformal Trisazine (TFTA) 4-amino-3,5-dimethyl-1,2,4-triazole	$C_4H_{12}N_6$	+28	$4HCHO + 3N_2H_4H_2O$ (Ref. 13)	

Note. Fuels 4–9 are hydrazine based.

Overall, these fuels serve the following purposes:

1. They are the source of C and H, which on combustion form simple gaseous molecules of CO_2 and H_2O and liberate heat.
2. They form complexes with the metal ions facilitating homogenous mixing of cations in solution.
3. They break down into components from which they are formed. These components in turn decompose to produce combustible gases like HNCO, NH_3 which ignite with NO_x.

Simple compounds such as urea and glycine are recognized as potential fuels. Compounds containing N–N bonds in their moieties are particularly found to assist the combustion better. Some important criteria that qualify an ideal fuel are

- Be water soluble.
- Have low ignition temperature ($<500°C$).
- Be compatible with metal nitrates, i.e., the combustion reaction should be controlled and smooth and not lead to explosion.
- Evolve large amounts of gases that are of low molecular weight and harmless during combustion.
- Yield no other residual mass except the oxide in question.
- Be readily available or easy to prepare.

A combustion synthesis reaction is influenced by the type of fuel and the fuel-to-oxidizer ratio. The exothermic temperature of the redox reaction (T_{ad}) varies from $1000°C$ to $1500°C$. Depending upon the fuel used and the type of metal ion involved, the nature of combustion differs from flaming (gas phase) to nonflaming (smouldering and heterogeneous) type. Flaming reactions could be attributed to the generation of gaseous products like nitrogen oxides (NO_x) by metal nitrates and HNCO, NH_3, CO, etc., generated by fuels like urea.

Interestingly, some of these fuels were found to be specific to a particular class of oxides. Urea for example is specific for alumina and related oxides. Similarly, CH is specific for zirconia and related oxides; ODH for Fe_2O_3 and ferrites; TFTA for TiO_2 and related oxides; glycine for chromium and related oxides, etc. The fuel specificity appears to be dictated by the metal–ligand complex formation, the thermodynamics of the reaction as well as thermal stability of the desired oxide formed.

Theoretically any redox mixture once ignited, undergoes combustion. All metal nitrates on pyrolysis yield corresponding metal oxides. The decomposition temperature of the metal nitrates is lowered by the addition of a fuel. So the choice of fuel is critical in deciding the exothermicity of the redox reaction between the metal nitrate and the fuel. Depending upon the exothermicity of the reaction, combustion is smoldering, flaming, or explosive. For example, aluminum nitrate–urea reaction is highly exothermic ($T_{ad} \sim 1500°C$) but is not explosive, probably due to the thermal insulating nature of the alumina formed. Whereas transition metal nitrate–urea reaction is violent. By changing the fuel from urea to CH or glycine, the combustion is much more controlled due to the complex formation of the metal ions with a ligand like CH.

3.4 A RECIPE FOR THE SYNTHESIS OF VARIOUS CLASSES OF OXIDES

SCS has been used to prepare a wide variety of oxide materials of desired composition and structure. This has been illustrated by reactions of metal nitrates with ODH, a unique fuel.[14]

Based on the valency concept, the reducing valency of ODH ($C_2H_6N_4O_2$) is +10. The oxidizing valencies of mono-, di-, tri-, and tetravalent metal nitrates are -5, -10, -15, and -20, respectively. Accordingly, for a divalent $M(NO_3)_2$–ODH mixture the equivalence ratio (Φ_e) is, $10/10 = 1.0$ and for a trivalent $M(NO_3)_3$–ODH mixture it is $15/10 = 1.5$. The mole ratios of metal nitrates and ODH become $1:1$ for the preparation of AO type oxides, $1:1.5$ for the preparation of A_2O_3 type oxides, $1:2:4$ for the preparation of AB_2O_4 type oxides, $2:1:4$ for the preparation of A_2BO_4 type oxides, and $1:1:3$ for the preparation of ABO_3 type oxides. Such mixtures on rapid heating ignite and burn completely without leaving any carbon residue. Assuming complete combustion for some of these oxides, theoretical equations can be written as follows:

AO:

$$A(NO_3)_2(aq) + C_2H_6N_4O_2(aq) \longrightarrow AO(s) + 2CO_2(g) \\ + 3N_2(g) + 3H_2O(g) \tag{10}$$

where A = Mg, Zn, Ni, Cu, . . .

A_2O_3:

$$2A(NO_3)_3(aq) + 3C_2H_6N_4O_2(aq) \longrightarrow A_2O_3(s) + 6CO_2(g)$$
$$+ 9N_2(g) + 9H_2O(g) \tag{11}$$

where A = Al, Cr, Fe, . . .

AB_2O_4 *(Spinel)*:

$$A(NO_3)_2(aq) + 2B(NO_3)_3(aq) + 4C_2H_6N_4O_2(aq) \longrightarrow AB_2O_4(s)$$
$$+ 8CO_2(g) + 12N_2(g) + 12H_2O(g) \tag{12}$$

where A = Mg, Mn, Co, Ni, Cu, Zn, . . . B = Al, Fe, Cr, . . .

A_2BO_4 *(K_2NiF_4 type)*:

$$2A(NO_3)_3(aq) + B(NO_3)_2(aq) + 4C_2H_6N_4O_2(aq) \longrightarrow A_2BO_4(s)$$
$$+ 8CO_2(g) + 12N_2(g) + 12H_2O(g) \tag{13}$$

where A = La, . . .   B = Mn, Co, Ni, Cu, Sr, . . .

ABO_3 *(Perovskite)*:

$$A(NO_3)_3(aq) + B(NO_3)_3(aq) + 3C_2H_6N_4O_2(aq) \longrightarrow ABO_3(s)$$
$$+ 6CO_2(g) + 9N_2(g) + 9H_2O(g) \tag{14}$$

where A = La, . . .   B = Al, Fe, Cr, . . .

$A_3B_5O_{12}$ *(Garnet)*:

$$3A(NO_3)_3(aq) + 5B(NO_3)_3(aq) + 12C_2H_6N_4O_2(aq)$$
$$\longrightarrow A_3B_5O_{12}(s) + 24CO_2(g) + 36N_2(g) + 36H_2O(g) \tag{15}$$

where A = Y, . . .   B = Al, Fe, . . .

$AB_{12}O_{19}$ *(Hexaferrites)*:

$$A(NO_3)_3(aq) + 12B(NO_3)_3(aq) + 19C_2H_6N_4O_2(aq)$$
$$\longrightarrow AB_{12}O_{19}(s) + 38CO_2(g) + 57N_2(g) + 57H_2O(g) \tag{16}$$

where A = Ba, Sr, . . . B = Al, Fe, . . .

It is interesting to note that the number of moles of ODH required for the combustion synthesis of oxides is equal to the number of oxygen atoms in the product, e.g., 4 mol of ODH for AB_2O_4 and A_2BO_4 type oxides and 3 mol for ABO_3 type oxides. This is true for any fuel with a reducing valency of $+10$ and metal nitrates as oxidizer.

The general scheme for the preparation of any oxide material is shown in Fig. 3.4.

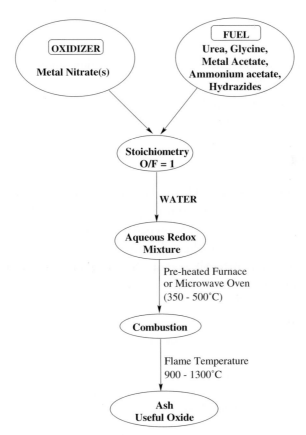

Fig. 3.4.   Flow diagram for preparation of oxide materials by solution combustion synthesis.

Caution-Some Safety Requirements. The combustion of redox mixtures if not controlled can lead to explosions. Hence, all reactions are to be carried out in open containers with minimum quantity of redox mixture, in fume cupboard with efficient exhaust. One should wear safety glasses and carry out the reactions under the protection of safety shield.

3.4.1 *Recipe for Nanomaterials*

It is possible to control the particle size of solid combustion product (ash) to nano-range by using suitable fuels that control the nature of combustion. This results in:

- Slow burning rate.
- Reduction in exothermicity (making the reaction smoldering instead of flaming, $T_{ad} < 1000°C$).
- Increase in number of gaseous products (water, nitrogen, carbon dioxide, etc.), which dissipate the heat.
- Linear combustion which appears to yield nano-oxides unlike volume combustion.

Hydrazide compounds like CH, ODH, glycine, ammonium acetate (AA), or metal acetate appear to serve this purpose. In this way a large number of nanocrystalline oxide materials are prepared and characterized by the choice of appropriate precursors in the redox mixture (Table 3.2). Formation of nanocrystalline oxide materials is confirmed by X-ray line broadening, surface area measurement, and TEM.

Table 3.2.   Nano-oxides prepared by solution combustion synthesis.

Oxides	Application	Fuel	Surface area $(m^2 g^{-1})$	Particle size (nm)
CeO_2	Polishing lenses	ODH	84	19
		CH	85	18
		GLY	30	15
$Ce_{1-x}Pr_xO_2$	Red pigment	ODH	85	18
ZrO_2	Coatings	CH	8	30
		GLY	17	29
$Ce_{1-x}Zr_xO_2$	Oxygen storage catalysts (OSC)	CH	80–120	18–27

(*Continued*)

Table 3.2. (*Continued*)

Oxides	Application	Fuel	Surface area $(m^2\,g^{-1})$	Particle size (nm)
$Ce_{1-x}M_xO_{2-\delta}$, $M^0/\alpha\text{-}Al_2O_3$ (M^0 = Cu, Ag, Au, Pd, and Pt)	Three-way automobile exhaust catalysts (TWC)	ODH	84	19[a]
$Ti_{1-x}M_xO_{2-\delta}$, M = Pt and Pd	Photocatalysts	GLY	156	12[a]
		CH	100	18[a]
		ODH	90	20[a]
$\gamma\text{-}Fe_2O_3$	Recording material, cancer detection and remediation	MDH	45	38
Ferrites (MFe_2O_4, $LnFeO_3/$ $Ln_3Fe_5O_{12}$)	Microwave applications, TV deflector tubes	ODH	50–140	17–90
$BaFe_{12}O_{19}$	Permanent magnets, recording tapes and discs	ODH	62	30–40
$ZrO_2\text{-}Al_2O_3$ (ZTA)	Toughened ceramic	CH	40–65	53–23
Eu^{3+}/Y_2O_3	Red phosphor	ODH	19	100[a]
Ni–YSZ	Anode material for SOFC	CH	29–32	13–20
Al–Borate ($9Al_2O_3\cdot2B_2O_3$)	Light weight ceramic	U	60	40
		AA	105	29
Eu^{2+}-doped $BaMgAl_{10}O_{17}$	Blue phosphor	U	10	160[a]
$Tb^{3+}/\,Ce^{3+}$-doped $MgCeAl_{11}O_{19}$	Green phosphor (CRT displays)	U	11	130[a]
$Cr^{3+}/\alpha\text{-}Al_2O_3$ (pink alumina)	Abrasive, pigment, synthetic gem (ruby), laser	U	8	220[a]
$Co^{2+}/\alpha\text{-}Al_2O_3$ (blue alumina)	Blue pigment	U	8	220[a]
$Co^{2+}/Mg_2B_2O_5$	Pink pigment	CH	8	320[a]
MAl_2O_4 (M=Mn, Co, Ni, and Cu)	Pigments	CH	43–83	53–23
$FeVO_4$	Laser host	ODH	30	20
$BaTiO_3$	Piezoelectric material	GLY	35	18–25
PZT		TFTA	19	50
$La(Sr)MnO_3$	Cathode material for SOFC	ODH	12–19	13–20
$La(Sr)FeO_3$	Heating element	ODH	21–54	28–67

[a] All the dopants (1–2 atom%) are of nanosize.

3.5 SALIENT FEATURES OF SOLUTION COMBUSTION METHOD

The exothermicity of the redox chemical reaction in SCS is used to produce useful materials. It has emerged as a viable technique for the preparation of advanced ceramics, catalysts, phosphors, pigments, composites, intermetallics, and nanomaterials as depicted in Fig. 3.5.

Fig. 3.5.   Nano-creations of fire. (1) $SrAl_2O_4$. (2) Rare earth phosphors. (3) Zirconia foam. (4) Pd/CeO_2-coated honeycomb monolith. (5) TEM of zirconia. (6) TEM of PZT. (7) Pink pigment ($Co/Mg_2B_2O_5$). (8) $NiFe_2O_4$.

The salient advantages of SCS process are

- It is an easy and fast process that uses relatively simple equipment.
- Composition, structure, homogeneity, and stoichiometry of the products can be controlled.
- Formation of high-purity products are ensured by this method.
- High exothermicity of the metal nitrate–fuel reaction permits incorporation of desired quantity of impurity ions or dopants in the oxide hosts to prepare

industrially useful materials like magnetic oxides (Co-doped Fe_2O_3), pigments (Co^{2+}/Al_2O_3, Cr^{3+}/Al_2O_3, $Ce_{1-x}Pr_xO_2$), and phosphors (Eu^{3+}/Y_2O_3 — red, Tb^{3+}/Y_2O_3 — green) as well as high-T_c cuprates, SOFC materials, and M/oxide catalysts ($Ce_{1-x}Pt_xO_2$) and tough materials like t-ZrO_2/Al_2O_3. The nanosize of the dopants (1–2 atom%) dramatically changes the properties of the host material.

- Stabilization of metastable phases (γ-Fe_2O_3, t-ZrO_2, anatase TiO_2, etc.) is possible by this method.
- Formation of products of virtually any size (micron to nano) and shape (spherical to hexagonal) can be achieved by this process.
- This method involves lower costs of preparation compared to conventional ceramic methods.
- It is economically attractive and easy to scale up.
- Uniform distribution of the dopants takes place throughout the host material due to the atomic mixing of the reactants in the initial solution.

References

1. Merzhanov AG, Combustion: New manifestation of an ancient process, in Rao CNR (ed.), *Chemistry of Advanced Materials*, Blackwell Scientific, Oxford, pp. 19–39, 1993.
2. Merzhanov AG (ed.), *SHS Research and Development; Handbook*, Institute of Structural Macrokinetics and Materials Science, Russian Academy of Sciences, Chernogolovka, Russia, 1999.
3. Kingsley JJ, Patil KC, A novel combustion process for the synthesis of fine particle α-alumina and related oxide materials, *Mater Lett* **6**: 427–432, 1988.
4. Sarner SF, *Propellant Chemistry*, Reinhold Publishing Corporation, New York, 1966.
5. Jain SR, Adiga KC, Pai Vernekar VR, A new approach to thermochemical calculations of condensed fuel-oxidiser mixtures, *Combust Flame* **40**: 71–79, 1981.
6. Adams GK, Parker WG, Wolfhard HG, Radical reactions of nitric oxide in flames, *Discussions Faraday Soc* **14**: 97–103, 1953.
7. Kiminami RHGA, Morelli MR, Folz DC, Clark DE, Microwave synthesis of alumina powders, *Am Ceram Soc Bull* **79**: 63–67, 2000.
8. Mohr EB, Brezinski JJ, Audrieth LF, Carbohydrazide, in Bailer Jr JC (ed.), *Inorganic Synthesis*, Vol. 4, McGraw Hill, New York, pp. 32–35, 1953.
9. Gran G, The use of oxalyl dihydrazide in a new reaction for spectrophotometric microdetermination of copper, *Anal Chim Acta* **14**: 150–156, 1956.

10. Ahmed AD, Mandal PK, Chaudhuri NR, Metal complexes of malonyl-dihydrazide, *J Inorg Nucl Chem* **28**: 2951–2959, 1966.
11. Mizzoni RH, Spoerri PE, Synthesis in the pyridazine and 3,6-dichloropyridazine, *J Am Chem Soc* **73**: 1873–1874, 1951.
12. Mashima M, The infrared absorption spectra of condensation products of formaldehyde with hydrazide, *Bull Chem Soc Jpn* **89**: 504–506, 1966.
13. Ainsworth C, Jones RG, Isomeric and nuclear substituted β-aminoethyl-1, 2,4 triazoles, *J Am Chem Soc* **77**: 621–624, 1955.
14. Suresh K, Patil KC, A recipe for an instant synthesis of fine particle oxide materials, in Rao KJ (ed.), *Perspectives in Solid State Chemistry*, Narosa Publishing House, New Delhi, pp. 376–388, 1995.

Chapter 4

Alumina and Related Oxide Materials

4.1 INTRODUCTION

Alumina (Al_2O_3) covers about 20% of the earth's crust and occurs both in amorphous-hydrated form ($Al_2O_3 \cdot xH_2O$) and crystalline forms like α-Al_2O_3, β-Al_2O_3, γ-Al_2O_3, and ζ-Al_2O_3. It is an extremely stable material with versatile properties including excellent mechanical strength with hardness of 9.0 on Mohs' scale, a high melting point (2041°C), and dimensional stability up to 1500°C, chemical inertness to most acids and alkalis and resistance to corrosion in hostile environments. Such unique properties coupled with its low cost makes high-purity alumina (up to 99.99% purity) the most commonly used material for engineering applications such as electronics, chemical processing, oil and gas processing, refractory, preparation of synthetic gems, wear and corrosion resistance, and a wide variety of other industrial processes. An important development that has hitherto taken place at the industrial floor is the gradual replacement of conventional metal or plastic materials with alumina-based ceramic materials. This has tremendously increased the demand for alumina-like materials or alumina-based materials.

Diverse applications of alumina are listed in Table 4.1 to emphasize the importance of purity levels involved in the use of alumina.[1] More than 130 million tons of bauxite is mined world wide each year and approximately 85% is used in the production of aluminum metal. The remaining is used in the fabrication of refractory materials and abrasives, alumina-based chemicals as fillers for plastics and tooth paste, activated alumina as a desiccant, filtering

Table 4.1.   Applications of various grades of alumina.

Purity levels	Applications
Ultra high-purity alumina (99.999%)	Polycrystalline alumina tubes for high-pressure sodium vapor lamps
High-Purity alumina (99.99%)	Alumina substrates, sliders for computer disc drives and other optical equipments, catalyst and catalyst carriers, cladding of control rods
Reactive alumina (99.8–99.9%)	Ignition electrodes, textile thread guides, bearing and piston plungers, cones for paper and pulp industry
Standard purity alumina (99.8%)	Sand blasting nozzles, seal faces for mechanical seals and pump components
Intermediate soda grade alumina (93%)	Spark plugs, fused bodies, water faucets discs, material handling liners, armor plating, thermal beads for heating pads

media, chromatographic column packing, and as an inert support for finely divided noble metals.

4.2 ALUMINA AND RELATED OXIDE MATERIALS

Alumina crystallizes with corundum structure (Fig. 4.1) having sixfold coordination with oxygen. The O-atoms are placed in HCP array but only two-third of the octahedral sites are occupied as shown in the motif. Strong Al–O–Al cross-linking takes place in three-dimensions in which AlO_6 octahedra share not only corners but also edges and faces as well.

These properties are reflected in the fact that mineral corundum is the second hardest substance known, is chemically inert and has high melting point ($>2000°C$). Various applications of alumina and related oxide materials are listed in Table 4.2.

The synthesis of alumina and related materials is achieved either by ceramic method (solid-state reaction) or by wet chemical methods like sol–gel, hydrothermal, co-precipitation, spray decomposition, and freeze drying. Solid state methods do not give fine oxides while wet chemical techniques yield fine oxides that are high value-added inorganic materials with controlled microstructure and properties. The manufacture of these products requires

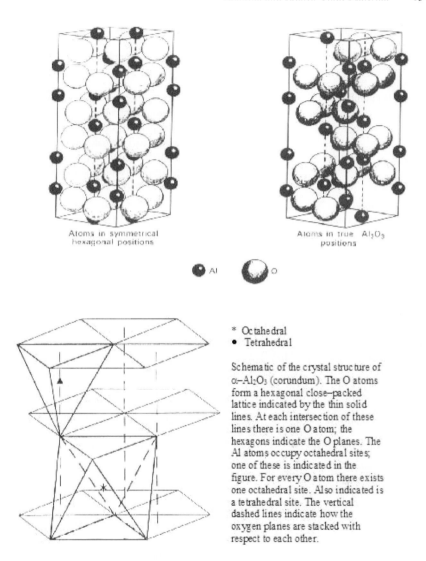

Fig. 4.1.   Corundum structure.

temperatures greater than 1000°C, long processing time, costly chemicals, or special equipments that need careful maintenance.

In the following section, preparation of α-alumina and related oxide materials by solution combustion method along with their properties have been described.

Table 4.2.   Applications of alumina and related oxide materials.

Materials	Applications
Alumina	
α-Al_2O_3 (alumina)	Refractory, abrasive, substrate for electronic packaging, cutting tools, spark plugs, cladding control rods of radioactive materials, fiber optic connectors, food machine parts, valves, pumps, fire retardants, smoke suppressors and filters
Doped alumina	
Cr^{3+}/α-Al_2O_3 (pink alumina)	Abrasive, pigment, synthetic gem (ruby), laser
Co^{2+}/α-Al_2O_3 (blue alumina)	Blue pigment
$Ni^{2+}/SrO\cdot6Al_2O_3$	Blue pigment
M^0/α-Al_2O_3 (M^0 = Pt, Pd, Ag)	Three-way automobile exhaust catalysts
t-ZrO_2/Al_2O_3 (ZTA)	Toughened ceramic, ceramic engine, scissors and knives
Aluminates	
$MgAl_2O_4$ and $ZnAl_2O_4$	Abrasives
$CoAl_2O_4$ and Co^{2+}-doped $MgAl_2O_4$	Blue pigments
$CaAl_2O_4$	Quick set cement
Orthoaluminates	
$LnAlO_3$ (Ln= lanthanide ions, La, Ce, Pr, Gd)	Laser hosts, perovskite substrate for high-T_c cuprate and CMR films, MHD generators
Cr^{3+}-doped $LaAlO_3$	Phosphor and pigments
Hexaaluminates	
$BaMgAl_{10}O_{17}$ and $MgLaAl_{11}O_{19}$	Yellow Phosphor
Eu^{2+} doped $BaMgAl_{10}O_{17}$	Blue phosphor
Tb^{3+} doped $MgCeAl_{11}O_{19}$	Green phosphor used in TV and CRT displays
Garnets	
$Y_3Al_5O_{12}$	Refractory material
Cr^{3+}-and Nd^{3+}-doped $Y_3Al_5O_{12}$	YAG laser
Minerals	
$3Al_2O_3\cdot2SiO_2$ (mullite)	Refractory, infrared window material
$Mg_2Al_4Si_5O_{18}$ (cordierite)	Electronic packaging, honey combs
Al_2TiO_5 (tialite)	Refractory material
Others	
$AlPO_4$ (aluminum phosphate)	Molecular sieves
$\beta-NaAl_{11}O_{17}$ (β-alumina)	Solid electrolyte used in batteries
$Al_{18}B_4O_{33}$ (aluminium borate)	Light weight ceramic

4.3 α-ALUMINA

α-Al_2O_3 is prepared by combustion of aqueous solutions containing stoichiometric quantities of aluminum nitrate and fuels like urea, DFH, HMT, GLY, ODH, and CH.[2–8]

All the reactions are flaming type ($T_{ad}\sim1500°C$) except when glycine is used. In the case of GLY and HMT fuels, stoichiometry of the reaction is achieved by the addition of extra oxidizer like NH_4NO_3/NH_4ClO_4. Theoretical equations for the combustion reaction of the redox mixtures to yield alumina can be written as follows:

$$2Al(NO_3)_3(aq) + 5\underset{U}{CH_4N_2O}\,(aq)$$

$$\overset{500°C}{\longrightarrow} Al_2O_3(s) + 5CO_2(g) + 8N_2(g) + 10H_2O(g)$$

$$(23 \text{ mol of gases/mol of } Al_2O_3) \qquad (1)$$

$$8Al(NO_3)_3(aq) + 15\underset{DFH}{C_2H_4N_2O_2}(aq)$$

$$\longrightarrow 4Al_2O_3(s) + 30CO_2(g) + 30H_2O(g) + 27N_2(g)$$

$$(\sim22 \text{ mol of gases/mol of } Al_2O_3) \qquad (2)$$

$$2Al(NO_3)_3(aq) + \underset{HMT}{C_6H_{12}N_4}(aq) + 3NH_4NO_3(aq)$$

$$\longrightarrow Al_2O_3(s) + 6CO_2(g) + 12H_2O(g) + 8N_2(g)$$

$$(26 \text{ mol of gases/mol of } Al_2O_3) \qquad (3)$$

$$2Al(NO_3)_3(aq) + 4\underset{GLY}{C_2H_5NO_2}(aq) + 3NH_4NO_3(aq)$$

$$\longrightarrow Al_2O_3(s) + 8CO_2(g) + 16H_2O(g) + 8N_2(g)$$

$$(32 \text{ mol of gases/mol of } Al_2O_3) \qquad (4)$$

$$2Al(NO_3)_3(aq) + 3\underset{ODH}{C_2H_6N_4O_2}(aq)$$

$$\longrightarrow Al_2O_3(s) + 6CO_2(g) + 9H_2O(g) + 9N_2(g)$$

$$(24 \text{ mol of gases/mol of } Al_2O_3) \qquad (5)$$

$$8Al(NO_3)_3(aq) + 15CH_6N_4O(aq)$$
$$\xrightarrow{CH} 4Al_2O_3(s) + 15CO_2(g) + 45H_2O(g) + 42N_2(g)$$

$$(25.5 \text{ mol of gases/mol of } Al_2O_3) \qquad (6)$$

Formation of α-Al_2O_3 is confirmed by its powder XRD pattern (Fig. 4.2). The TEM of α-Al_2O_3 formed by the combustion process using urea as fuel shows nearly hexagonal platelet particles with sizes ranging from 0.2 to 0.8 μm (Fig. 4.3).

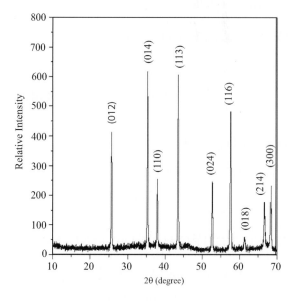

Fig. 4.2.   XRD pattern of α-Al_2O_3.

The particulate properties of various alumina obtained by using different fuels have been summarized in Table 4.3.

Al_2O_3 powder prepared by CH fuel has the highest BET surface area of 65 m^2 g^{-1}. Also, α-Al_2O_3 prepared by CH and GLY give nanosize particles. Both CH-derived alumina and ODH-derived alumina could be sintered to 85–96% theoretical density at 1500°C. With the addition of 3 mol% yttria, alumina could be sintered to 94% theoretical density at 1650°C. Micrographs of alumina and yttria–alumina sintered at various temperatures are shown in Fig. 4.4. They show a uniform grain size of 1.5 μm.

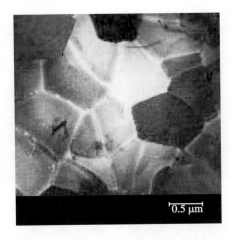

Fig. 4.3.   TEM of hexagonal α-Al$_2$O$_3$.

Table 4.3.   Compositions of redox mixtures using various fuels for as-Synthesized α-Al$_2$O$_3$ along with particulate properties.

Composition of the redox mixtures	Nature of combustion	Powder density (g cm^{-3})	Surface area (m^2 g^{-1})	Particle size[a] (nm)	Reference
20 g Al(NO$_3$)$_3$·9H$_2$O + 8 g urea	Flaming	3.21	8.30	220	2
Microwave: 20 g Al(NO$_3$)$_3$·9H$_2$O + 8 g urea	Flaming	3.78	25	60	3
20 g Al(NO$_3$)$_3$·9H$_2$O + 8.8 g DFH	Flaming	2.40	30	83	4
11.7 g Al(NO$_3$)$_3$·9H$_2$O + 1.863 g HMT + 0.586 g NH$_4$ClO$_4$	Flaming	3.93	42	36	5
5 g Al(NO$_3$)$_3$·9H$_2$O + 2.5 g GLY + 6.4 g NH$_4$NO$_3$	Smoldering	3.10	14	55	6
10 g Al(NO$_3$)$_3$·9H$_2$O + 4.6 g ODH	Flaming	3.12	12.2	160	7
7 g Al(NO$_3$)$_3$·9H$_2$O + 2.7 g CH	Flaming	2.6	65	360	8

[a] From surface area.

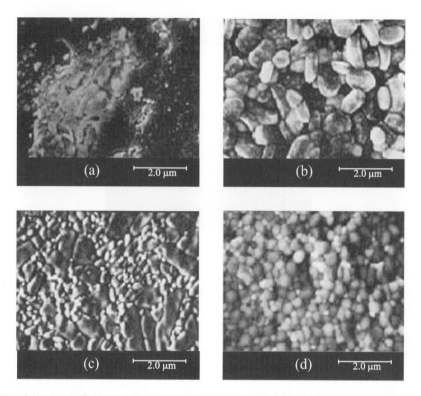

Fig. 4.4.   SEM of: (a) α-Al$_2$O$_3$ sintered at 1500°C, (b) Yttria–Al$_2$O$_3$ sintered at 1500°C, (c) α-Al$_2$O$_3$ sintered at 1600°C, and (d) Yttria–Al$_2$O$_3$ sintered at 1600°C.

4.4 METAL ALUMINATES (MAl$_2$O$_4$)

Metal aluminates (MAl$_2$O$_4$) crystallize with mineral spinel (MgAl$_2$O$_4$) structure. The basic structure is derived from cubic close packed spheres with a doubled face-centered cubic unit cell having eight formula units, i.e., M$_8$Al$_{16}$O$_{32}$ containing 32 octahedral holes and 64 tetrahedral holes. In the spinel structure 50% of the octahedral holes are designated as Al positions and one-eighth of the tetrahedral holes as M sites (Fig. 4.5).

Metal aluminates, MAl$_2$O$_4$ (where M = Mg, Ca, Sr, Ba, Mn, Fe, Co, Ni, Cu, Zn) with the mineral spinel structure are prepared by the combustion of aqueous solutions containing corresponding metal nitrates, aluminum nitrate, and fuels like urea, DFH, HMT, GLY, ODH, and CH.[2,4–7,9,10]

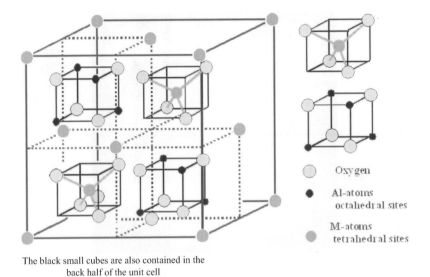

The black small cubes are also contained in the
back half of the unit cell

Fig. 4.5. AB_2O_4 spinel structure.

Interestingly aluminum nitrate–urea mixture, which undergoes combustion
with a flame ($T_{ad} \sim 1500°C$) is ideal to prepare nontransition metal aluminates
like MAl_2O_4, M = Mg, Ca, Ba, Sr, and Zn, etc., that are highly crystalline
and voluminous, e.g., $SrAl_2O_4$ (Fig. 4.6). However, this process is not suitable
to prepare transition metal aluminates like MAl_2O_4, M = Fe, Co, Ni, and
Cu since the metal aluminates formed are thermally unstable at that temperature. Fuels, like carbohydrazide or ODH were found to be suitable for the
preparation of these aluminates. The combustion of aluminum nitrate, metal
nitrate–CH/ODH is nonflaming and yields nanocrystalline MAl_2O_4, where
M = Mn, Co, Ni, Cu.

Typical theoretical equations assuming complete combustion for the preparation of various MAl_2O_4 are given below.

$$3M(NO_3)_2(aq) + 6Al(NO_3)_3(aq) + 20\underset{\text{Urea}}{CH_4N_2O}(aq)$$

$$\longrightarrow 3MAl_2O_4(s) + 20CO_2(g) + 40H_2O(g) + 32N_2(g)$$

$$(\sim 31 \text{ mol of gases/mol of } MAl_2O_4)$$

M = Mg, Ca, Sr, Ba and Zn   (7)

Fig. 4.6. As-prepared $SrAl_2O_4$.

$$M(NO_3)_2(aq) + 2Al(NO_3)_3(aq) + 5CH_6N_4O(aq)$$
$$\xrightarrow{CH} MAl_2O_4(s) + 5CO_2(g) + 15H_2O(g) + 14N_2(g)$$
$$(34 \text{ mol of gases/mol of } MAl_2O_4)$$

M = Fe, Co, Ni and Cu   (8)

$$M(NO_3)_2(aq) + 2Al(NO_3)_3(aq) + 4C_2H_6N_4O_2(aq)$$
$$\xrightarrow{ODH} MAl_2O_4(s) + 8CO_2(g) + 12N_2(g) + 12H_2O(g)$$
$$(32 \text{ mol of gases/mol of } MAl_2O_4)$$

M = Mg, Mn, Fe, Co, Ni, Cu and Zn   (9)

The preparation of zinc aluminate ($ZnAl_2O_4$), nickel aluminate ($NiAl_2O_4$), and copper aluminate ($CuAl_2O_4$) by the combustion of corresponding metal nitrate, aluminum nitrate with urea, CH, and ODH are given below as representatives.

Zinc aluminate is prepared by the combustion of an aqueous redox mixture containing 3.97 g $Zn(NO_3)_2 \cdot 6H_2O$, 10 g $Al(NO_3)_3 \cdot 9H_2O$, and 5.35 g urea.

The combustion reaction is flaming type, similar to the combustion reaction observed in case of α-Al_2O_3 preparation by urea. Nickel aluminate is prepared by dissolving 3.88 g of $Ni(NO_3)_2\cdot4H_2O$ and 10 g of $Al(NO_3)_3\cdot9H_2O$ in a minimum quantity of water along with 4.5 g of CH in a Pyrex dish. The solution boils and undergoes smooth deflagration with enormous swelling, producing light blue colored foam.

Copper aluminate is prepared by the combustion of 3.94 g of $Cu(NO_3)_2\cdot6H_2O$ and 10 g of $Al(NO_3)_3\cdot9H_2O$ with 6.3 g of ODH. The reaction is flaming and yields a greenish brown product with large amounts of gases.[7–9]

The as-formed $ZnAl_2O_4$ shows the typical XRD pattern of spinel (Fig. 4.7). Transition metal aluminates prepared by CH fuel are voluminous and crystalline whereas those prepared by urea and ODH are less voluminous, with comparatively low surface area (Fig. 4.8). The crystallite sizes calculated from X-ray line broadening are nanosize.

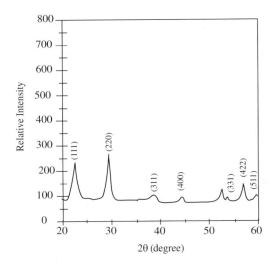

Fig. 4.7.   XRD pattern of $ZnAl_2O_4$.

The properties like density, particle size, and surface area of MAl_2O_4 (M = Zn, Mg, Ni, Co, Cu, Mn, Ca, Sr, and Ba) obtained by solution combustion method using various fuels are summarized in Table 4.4.

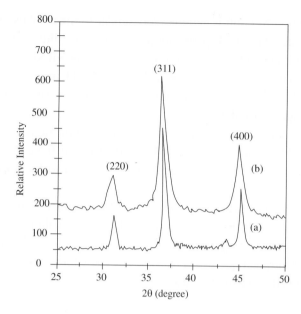

Fig. 4.8.   XRD pattern of $NiAl_2O_4$ prepared using (a) oxalyldihydrazide and (b) carbohydrazide fuels.

Metal aluminates (MAl_2O_4, M = Mg, Mn, Zn, Ca, Ba, etc.) are also prepared by using a mixture of metal acetates, aluminum nitrate and fuels like urea, HMT, CH, ODH, and GLY.[10] The reaction for metal acetate–nitrate mixture using ODH as the fuel assuming complete combustion may be represented by the following equation:

$$M(CH_3COO)_2(aq) + 2Al(NO_3)_3(aq)$$
$$+ 2NH_4NO_3(aq) + \underset{GLY}{2C_2H_5NO_2(aq)}$$
$$\longrightarrow MAl_2O_4(s) + 8CO_2(g) + 12H_2O(g) + 6N_2(g)$$
$$(26\ mol\ of\ gases/mol\ of\ MAl_2O_4) \quad (10)$$

In the acetate–fuel mixture the fire is of short duration and the combustion is more controlled than the metal nitrate–urea/carbohydrazide mixtures. Compared to the metal nitrate in the redox mixture, combustion with the acetate redox mixture is milder and yields nanosize metal aluminate spinels

Table 4.4.   Compositions of redox mixtures and particulate properties of metal aluminates.

Composition of the redox mixtures	Product and color[a]	Powder density $(g\,cm^{-3})$	Surface area $(m^2\,g^{-1})$	Particle size[b] (nm)
Urea (5.35 g)				
$Mg(NO_3)_2 \cdot 6H_2O$ (3.42 g) + A	$MgAl_2O_4$ (white)	3.00	21.80	100
$Ca(NO_3)_2 \cdot 4H_2O$ (3.15 g) + A	$CaAl_2O_4$ (white)	2.48	1.25	192
$Sr(NO_3)_2 \cdot 4H_2O$ (3.78 g) + A	$SrAl_2O_4$ (white)	2.88	2.41	86
$Ba(NO_3)_2$ (3.48 g) + A	$BaAl_2O_4$ (white)	3.20	2.01	93
$Zn(NO_3)_2 \cdot 6H_2O$ (3.97 g) + A	$ZnAl_2O_4$ (white)	3.60	8.70	19
Carbohydrazide (4.5 g)				
$Mn(NO_3)_2 \cdot 4H_2O$ (3.35 g) + A	$MnAl_2O_4$ (brown)	2.629	43.20	53
$Co(NO_3)_2 \cdot 6H_2O$ (3.88 g) + A	$CoAl_2O_4$ (thenard's blue)	3.262	58.30	32
$Ni(NO_3)_2 \cdot 6H_2O$ (3.88 g) + A	$NiAl_2O_4$ (sky-blue)	3.091	83.40	23
$Cu(NO_3)_2 \cdot 3H_2O$ (3.22 g) + A	$CuAl_2O_4$ (greenish brown)	3.710	59.80	27
Oxalyl dihydrazide (6.3 g)				
$Mg(NO_3)_2 \cdot 6H_2O$ (3.42 g) + A	$MgAl_2O_4$ (white)	3.12	22.40	90
$Mn(NO_3)_2 \cdot 4H_2O$ (3.34 g) + A	$MnAl_2O_4$ (brown)	2.83	10.80	210
$Co(NO_3)_2 \cdot 6H_2O$ (3.88 g) + A	$CoAl_2O_4$ (thenard's blue)	3.34	17.20	100
$Ni(NO_3)_2 \cdot 6H_2O$ (3.88 g) + A	$NiAl_2O_4$ (sky-blue)	3.13	11.40	170
$Cu(NO_3)_2 \cdot 3H_2O$ (3.22 g) + A	$CuAl_2O_4$ (greenish brown)	3.65	25.0	70
$Zn(NO_3)_2 \cdot 6H_2O$ (3.97 g) + A	$ZnAl_2O_4$ (white)	3.71	16.8	100

A = $Al(NO_3)_3 \cdot 9H_2O$ (10 g).
[a] Obtained at 500°C under normal atmospheric pressure.
[b] From surface area.

with large surface area. This has been suitably illustrated by the comparison of properties of $MnAl_2O_4$ derived from various fuels (Table 4.5).

4.5 RARE EARTH ORTHOALUMINATES (LnAlO$_3$)

Rare earth orthoaluminates with the general formula, $LnAlO_3$ where Ln = La, Pr, Nd, Sm, Eu, Gd, Tb, and Dy are prepared by solution combustion process using corresponding rare earth metal nitrate, aluminum nitrate, and urea/CH/ODH.[11] Assuming complete combustion, the theoretical equation

Table 4.5.   Particulate properties of manganese aluminate.

Oxide product	Powder density $(g\,cm^{-3})$	Surface area $(m^2\,g^{-1})$	Particle size[a] (nm)
$MnAl_2O_4$ (U)	4.3	180	—
$MnAl_2O_4$ (CH)	2.0 (2.6)	40 (43)	65 (53)
$MnAl_2O_4$ (ODH)	3.6 (2.8)	65 (53)	32 (210)
$MnAl_2O_4$ (HMT)	5.6	178	47
$MnAl_2O_4$ (GLY)	3.9	124	15

Values in parentheses obtained by using metal nitrate.
[a] From TEM.

for the formation of $LnAlO_3$ can be represented as follows:

$$Ln(NO_3)_3(aq) + Al(NO_3)_3(aq) + 5CH_4N_2O(aq)$$
$$\underset{Urea}{}$$
$$\longrightarrow LnAlO_3(s) + 5CO_2(g) + 10H_2O(g) + 8N_2(g)$$
$$(23\ mol\ of\ gases/mol\ of\ LaAlO_3) \qquad (11)$$

$$4Ln(NO_3)_3(aq) + 4Al(NO_3)_3(aq) + 15CH_6N_4O(aq)$$
$$\underset{CH}{}$$
$$\longrightarrow 4LnAlO_3(s) + 15CO_2(g) + 45H_2O(g) + 42N_2(g)$$
$$(25.5\ mol\ of\ gases/mol\ of\ LaAlO_3) \qquad (12)$$

$$Ln(NO_3)_3(aq) + Al(NO_3)_3(aq) + 3C_2H_6N_4O_2(aq)$$
$$\underset{ODH}{}$$
$$\longrightarrow LnAlO_3(s) + 6CO_2(g) + 9H_2O(g) + 9N_2(g)$$
$$(24\ mol\ of\ gases/mol\ of\ LaAlO_3) \qquad (13)$$

The formation of single-phase product (rare earth orthoaluminate) is confirmed by characteristic X-ray powder diffraction patterns (Fig. 4.9).

The compositions of redox mixtures used for the preparation of rare earth aluminates and their particulate properties are listed in Table 4.6.

4.6 GARNETS

Yttrium aluminum garnet (YAG) $(Y_3Al_5O_{12})$ has the cubic garnet structure, which is represented by the formula $C_3A_2D_3O_{12}$ where all the Y ions occupy

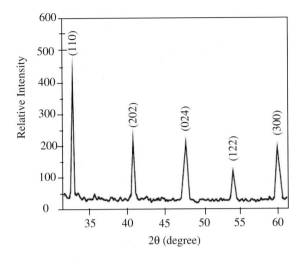

Fig. 4.9. XRD pattern of LnAlO$_3$.

dodecahedral C sites and the Al ions occupy both tetrahedral D and trigonally distorted octahedral A sites. This structure has eight formula units per cell and belongs to 1a3d space group. YAG is a ceramic with interesting mechanical and optical properties. It is a refractory oxide that does not easily damage under high irradiance with electron beam. YAG is a good host to chromium and rare earth ions to form solid-state laser material.

YAG is prepared by solution combustion process using stoichiometric amounts of yttrium nitrate, aluminum nitrate, and urea/CH/ODH as fuels.[7,11] Theoretical equations for the formation of Y$_3$Al$_5$O$_{12}$ assuming complete combustion can be represented as follows:

$$3Y(NO_3)_3(aq) + 5Al(NO_3)_3(aq) + 20 \underset{\text{Urea}}{CH_4N_2O(aq)}$$

$$\longrightarrow Y_3Al_5O_{12}(s) + 20CO_2(g) + 40H_2O(g) + 32N_2(g)$$

$$(92 \text{ mol of gases/mol of } Y_3Al_5O_{12}) \quad (14)$$

Table 4.6.   Compositions of redox mixtures and particulate properties of rare earth orthoaluminates.

No.	Composition of redox mixtures	Product[a]	Powder density $(g\,cm^{-3})$	Surface area $(m^2\,g^{-1})$	Particle Size[b] (nm)
Urea (8.00 g)					
1.	La(NO$_3$)$_3$·6H$_2$O (11.54 g) + A	LaAlO$_3$	4.27	3	460
2.	Pr(NO$_3$)$_3$·6H$_2$O (11.59 g) + A	PrAlO$_3$	4.30	3.4	410
3.	Nd(NO$_3$)$_3$·6H$_2$O (11.68 g) + A	NdAlO$_3$	4.50	3.0	440
4.	Sm(NO$_3$)$_3$·6H$_2$O (11.84 g) +A	SmAlO$_3$	4.60	3.1	420
5.	Dy(NO$_3$)$_3$·5H$_2$O (11.69 g) + A	DyAlO$_3$	4.75	3.5	360
Carbohydrazide (7.2 g)					
1.	La(NO$_3$)$_3$·6H$_2$O (11.54 g) + A	LaAlO$_3$	4.89	3.5	350
2.	Pr(NO$_3$)$_3$·6H$_2$O (11.59 g) + A	PrAlO$_3$	4.91	3.4	360
3.	Nd(NO$_3$)$_3$·6H$_2$O (11.68 g) + A	NdAlO$_3$	4.93	3.2	380
4.	Sm(NO$_3$)$_3$·6H$_2$O (11.84 g) + A	SmAlO$_3$	4.92	3.1	390
5.	Dy(NO$_3$)$_3$·5H$_2$O (11.69 g) + A	DyAlO$_3$	4.82	4	310
Oxalyl dihydrazide (6.3 g)					
1.	La(NO$_3$)$_3$·6H$_2$O (11.54 g) + A	LaAlO$_3$	4.27	5.4	260
2.	Pr(NO$_3$)$_3$·6H$_2$O (11.59 g) + A	PrAlO$_3$	4.80	5.3	240
3.	Nd(NO$_3$)$_3$·6H$_2$O (11.68 g) + A	NdAlO$_3$	4.62	21.6	60
4.	Sm(NO$_3$)$_3$·6H$_2$O (11.84 g) + A	SmAlO$_3$	4.93	6.5	190
5.	Dy(NO$_3$)$_3$·5H$_2$O (11.69 g) + A	DyAlO$_3$	4.75	8.2	150
6.	Y(NO$_3$)$_3$·6H$_2$O (11.54 g) + A	YAlO$_3$	3.86	6.4	240

A = Al(NO$_3$)$_3$·9H$_2$O (10 g).
[a] Obtained at 500°C under normal atmospheric pressure.
[b] From surface area.

$$3Y(NO_3)_3(aq) + 5Al(NO_3)_3(aq) + 15CH_6N_4O(aq)$$
$$\xrightarrow{\text{CH}} Y_3Al_5O_{12}(s) + 15CO_2(g) + 45H_2O(g) + 42N_2(g)$$
$$(102 \text{ mol of gases/mol of } Y_3Al_5O_{12}) \quad (15)$$

$$3Y(NO_3)_3(aq) + 5Al(NO_3)_3(aq) + 12C_2H_6N_4O_2(aq)$$
$$\xrightarrow{\text{ODH}} Y_3Al_5O_{12}(s) + 24CO_2(g) + 36H_2O(g) + 36N_2(g)$$
$$(96 \text{ mol of gases/mol of } Y_3Al_5O_{12}) \quad (16)$$

The typical XRD pattern of the yttrium garnet is shown in Fig. 4.10. Lattice constants calculated from the XRD patterns are in good agreement with the literature $a = b = c = 1.198$ nm ± 0.002 nm. The powder density of YAG prepared both by urea and CH process are in the range 70–80% of the theoretical density (Table 4.7).

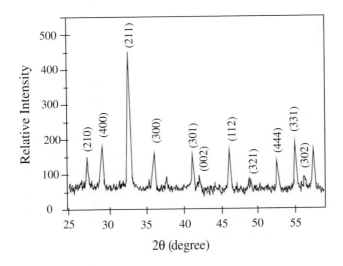

Fig. 4.10.   XRD pattern of yttrium aluminum garnet ($Y_3Al_5O_{12}$).

Table 4.7.   Compositions of redox mixtures and particulate properties of yttrium aluminum garnet.

Composition of the redox mixtures	Powder density $(g\,cm^{-3})$	Surface area $(m^2\,g^{-1})$	Particle size[a] (nm)
Urea (6.4 g) + A + B	3.86	7.3	210
Carbohydrazide (5.76 g) + A + B	3.81	3.3	470
Oxalyl dihydrazide (6.3 g) + A + B	3.83	6.3	250

A = $Al(NO_3)_3 \cdot 9H_2O$ (10 g); B = $Y(NO_3)_3 \cdot 6H_2O$ (6.12 g).
[a] From surface area.

4.7 ALUMINUM BORATE

Aluminum borate ($Al_{18}B_4O_{33}$) is important due to its attractive physical properties like low density ($2.68\,g\,cm^{-3}$) and thermal expansion coefficient ($4.2 \times 10^{-6}\,K^{-1}$), moderate strength ($250\,MPa$), and high melting point ($1950°C$). The average coefficient of thermal expansion of aluminum borate is smaller than that of α-alumina and is quite similar to that of mullite ($3Al_2O_3 \cdot 2SiO_2$). The single-crystal X-ray structure of the compound revealed that it has structure similar to mullite and is composed of AlO_6 octahedra, AlO_4 tetrahedra, B_2O_3 triangles, and five oxygen-coordinated aluminum atoms.

Fine-particle aluminum borate, ($Al_{18}B_4O_{33}$), ($9Al_2O_3 \cdot 2B_2O_3$), is prepared by the combustion of aqueous solution containing stoichiometric amounts of aluminum nitrate, boric acid, and urea/diformyl hydrazide/ammonium acetate in the mole ratio of 18:4:45 or 18:4:33.75 or 18:4:24.547.[12] In a typical experiment, an aqueous solution of $Al(NO_3)_3$ (20 g), H_3BO_3 (0.73 g), and DFH (8.79 g) when rapidly heated in a muffle furnace at $500°C$ ignites to yield voluminous powder. The formation of aluminum borate by SCS assuming complete combustion can be represented by the following reaction:

$$18Al(NO_3)_3(aq) + 4H_3BO_3(aq) + 45CH_4N_2O(aq)$$
$$\xrightarrow{500°C} 9Al_2O_3 \cdot 2B_2O_3(s) + 45CO_2(g) + 72N_2(g) + 96H_2O(g)$$
$$(213\,mol\,of\,gases/mol\,of\,Al_{18}B_4O_{33}) \qquad (17)$$

Alumina and B_2O_3 formed *in situ* during combustion react metathetically, forming aluminum borate. When the process uses urea and DFH, the combustion is flaming type, while the ammonium acetate process is found to be smoldering type (without flame). DFH and urea process-derived aluminum borate shows a crystalline nature. However, the aluminum borate prepared by the ammonium acetate process is X-ray amorphous. Calcination of the amorphous aluminum borate powder at $1050°C$ gives a single-phase crystalline aluminum borate. The formation of single-phase crystalline aluminum borate by combustion method is confirmed by powder XRD (Fig. 4.11).

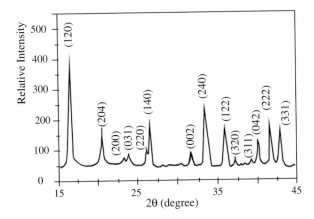

Fig. 4.11. XRD pattern of aluminum borate.

It is noted that urea-derived aluminum borate shows the presence of α-alumina, whereas the DFH process yields weakly crystalline aluminum borate. The formation of α-alumina in the urea process could be due to the decomposition of $Al_{18}B_4O_{33}$ at such high temperatures. However, a fuel lean mixture (O/F $\leq$ 1) of aluminum nitrate, boric acid and urea is found to yield single-phase crystalline orthorhombic $Al_{18}B_4O_{33}$. The formation of single-phase crystalline aluminum borate is probably due to the lowering of flame temperature of the fuel lean mixture, thereby preventing the decomposition of $Al_{18}B_4O_{33}$ into α-alumina. Controlled combustion of redox mixtures appears to prevent the decomposition of $Al_{18}B_4O_{33}$, back into α-Al_2O_3 and B_2O_3.

The particulate properties of aluminum borate prepared from all the three fuels are summarized in Table 4.8. The powder densities of the combustion products are 75–90% of theoretical values. The high surface area value of ammonium acetate-derived aluminum borate could be attributed to the smoldering (flameless) type of combustion and evolution of large amount of cold gases during combustion which dissipate the heat, thereby inhibiting sintering of the combustion product.

The particulate properties of combustion-derived aluminum borate vary depending upon the fuel used. The marked difference is attributed to the choice

Table 4.8.   Particulate properties of combustion-derived aluminum borate.

Fuel-based process	Powder density (g cm^{-3})	Surface area (m^2 g^{-1})	Particle size[a] (nm)
Urea process	2.50	60	40
DFH process	2.45	15	160
AA process	2.027	105	29

[a] From surface area.

of fuel which alter the energetics of the combustion reaction, and eventually, the properties of the combustion product.

4.8 TIALITE (β-Al$_2$TiO$_5$)

Aluminum titanate (tialite) exists in two forms, α and β. The α-form is stable at high temperatures of 1820°C, while the β form is stable at 1320–1400°C. β-Tialite exhibits excellent thermal shock resistance property that makes it an attractive refractory material for applications in metallurgy and automotive engineering.

Tialite is prepared by the combustion of aqueous solutions containing stoichiometric amounts of titanyl nitrate, aluminum nitrate, and urea/CH fuels. The Petri dish containing the solution when rapidly heated in a muffle furnace maintained at 500°C, boils, froths, and ignites to burn with a flame (1300 ± 50°C) yielding voluminous β-tialite powder.[13] Assuming complete combustion, the theoretical equation for the formation of tialite may be written as follows:

$$3TiO(NO_3)_2(aq) + 6Al(NO_3)_3(aq) + \underset{\text{Urea}}{20CH_4N_2O(aq)}$$
$$\longrightarrow 3Al_2TiO_5(s) + 20CO_2(g) + 40H_2O(g) + 32N_2(g)$$
$$(\sim31 \text{ mol of gases/mol of } Al_2TiO_5) \qquad (18)$$

$$TiO(NO_3)_2(aq) + 2Al(NO_3)_3(aq) + \underset{\text{CH}}{5CH_6N_4O(aq)}$$
$$\longrightarrow Al_2TiO_5(s) + 5CO_2(g) + 15H_2O(g) + 14N_2(g)$$
$$(34 \text{ mol of gases/mol of } Al_2TiO_5) \qquad (19)$$

The formation of crystalline tialite phase is confirmed by powder XRD. The ratios of the various phases present in tialite are calculated from the integral intensities of the major peaks (110) for β-Al$_2$TiO$_5$, (111) for rutile, and (113) for corundum. From the XRD analysis it is found that the CH process yields powders containing about 50 mol% of β-Al$_2$TiO$_5$, 25 mol% of rutile and 25 mol% of corundum, whereas the urea process gives 20 mol% β-Al$_2$TiO$_5$, 40 mol% rutile, and 40 mol% corundum. When the as-prepared powders are calcined at 1300°C for 30 min, a single-phase tialite is obtained (Fig. 4.12). The tialite prepared by the CH process has a higher density and higher surface area, which may be due to the higher amounts of gases evolved in the CH process (34 mol). Tialite powder derived by solution combustion processes is sinterable and the pellets sintered at 1300°C for 1 h achieved a density of 90–91% theoretical density and magnesia-doped Al$_2$TiO$_5$ achieved 94% theoretical density (Fig. 4.13).

The stability of Al$_2$TiO$_5$ is increased by doping with Mg, Si, and Zr oxides. Tialite and doped tialite (5 wt.% MgO) are prepared using urea and CH fuels. The composition of the redox mixtures used for combustion and the particulate properties of tialite formed are summarized in Table 4.9.

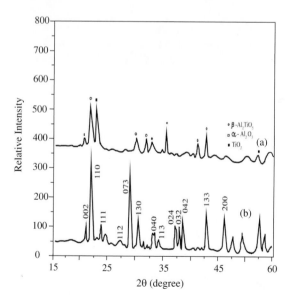

Fig. 4.12.   XRD pattern of β-Al$_2$TiO$_5$ (a) as-prepared and (b) calcined at 1300°C for 30 min.

Fig. 4.13.   SEM of tialite (a) CH process and (b) MgO-doped tialite, sintered at 1300°C for 1 h.

Table 4.9.   Compositions of the redox mixtures and particulate properties of tialite.

Composition of redox mixtures	Powder density $(g\,cm^{-3})$	Surface area $(m^2\,g^{-1})$	Particle size[a] (nm)
A + B + 10.60 g CH	1.63	25.13	150
A + B + C + 11.92 g CH	1.59	23.22	160
A + B + 9.23 g urea	1.28	19.45	240
A + B + C + 10.45 g urea	1.19	20.14	250

A—9.64 g $TiO(NO_3)_2$; B—9.25 g $Al(NO_3)_3 \cdot 9H_2O$; C—0.99 g $Mg(NO_3)_2 \cdot 6H_2O$; CH—carbohydrazide.
[a] From surface area.

4.9 ALUMINUM PHOSPHATE

Aluminum phosphate ($AlPO_4$) is widely used in the manufacture of optical glasses, enamels, and anticorrosion powder coatings. Due to its microporous nature it is used as molecular sieves and forms a good support for catalysts. It exists in several polymorphs.

$$\text{berlinite} \xrightarrow{820°C} \text{tridymite} \xrightarrow{1010°C} \text{crystobalite}$$

Aluminum phosphate is prepared by solution combustion synthesis using 5 g aluminum nitrate and 1.76 g diammonium hydrogen phosphate. These salts are dissolved in 40 mL of water in a cylindrical Pyrex dish to form a gel of amorphous hydrated aluminum phosphate. The gel is dissolved by adding a minimum amount of dilute HNO_3. To the homogeneous solution obtained 3.2 g ammonium nitrate and 2.25 g CH are added to form the combustion redox mixture. The dish containing the solution is introduced into a muffle furnace maintained at $400 \pm 10°C$. Initially the solution undergoes dehydration followed by decomposition, with the evolution of large amounts of gases. It then froths, ignites, and burns with a flame to yield voluminous aluminum phosphate powder.[14] The formation of aluminum phosphate by solution combustion assuming complete combustion may be explained by the following reaction sequence:

$$Al(NO_3)_3 + (NH_4)_2HPO_4$$
$$\xrightarrow{aqueous} AlPO_4 \cdot xH_2O + 2NH_4NO_3 + HNO_3 \tag{20}$$

$$AlPO_4 \cdot xH_2O(aq) + \underset{CH}{CH_6N_4O(aq)} + 4NH_4NO_3(aq)$$
$$\xrightarrow{400°C} AlPO_4(s) + CO_2(g) + (x+11)H_2O(g) + 6N_2(g)$$
$$(\sim 18 \text{ mol of gases-per mol of } AlPO_4)$$
$$\tag{21}$$

Combustion-derived $AlPO_4$ does not show the high-temperature polymorph crystobalite, instead it shows the tridymite phase as observed in the XRD (Fig. 4.14). The lattice constants $a = 1.73796$, $b = 1.43151$, $c = 0.98940$ nm confirm the formation of this phase.[14] Stabilization of the tridymite form of the $AlPO_4$ may be due to the presence of phosphate in the redox mixture, which is a well-known fire retardant. The as-formed aluminum phosphate powder has a density of 2.3 g cm^{-3}, surface area of 78 m^2 g^{-1}, and particle size of 30 nm.

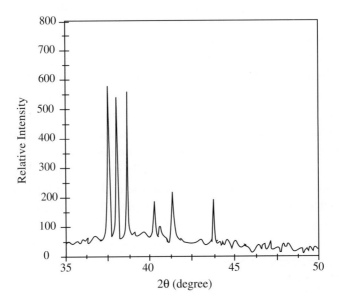

Fig. 4.14.   XRD pattern of $AlPO_4$.

4.10 ALUMINA COMPOSITES

Interestingly, metathetically formed alumina by solution combustion of aluminum nitrate–fuel (U, DFH, CH, etc.) is chemically highly reactive. This is seen when composite materials like mullite and cordierite are formed on combustion with respective metal oxides. Similarly, it is also possible to prepare SiAlON by the metathetical reaction of *in situ* formed alumina and fumed alumina with silicon nitride.

4.10.1 *Al$_2$O$_3$·SiO$_2$ System: Mullite*

Mullite (3Al$_2$O$_3$·2SiO$_2$) and cordierite (Mg$_2$Al$_4$Si$_5$O$_{18}$) are the two most stable phases in the alumina–silica binary system. These are promising materials for electronic packaging due to their low dielectric constant and coefficient of thermal expansion compared to alumina.

Mullite is prepared by the combustion of an aqueous heterogeneous mixture containing stoichiometric quantities of Al(NO$_3$)$_3$, silica fume (surface area = 200 m^2 g^{-1}), and urea. The composition of the redox mixtures are given in Table 4.10. The mixture, when rapidly heated at 500 ± 10°C in a muffle furnace boils, foams, and ignites to burn with a flame (temperature 1275 ± 25°C) to yield voluminous, foamy, weakly crystalline mullite.[15]

Table 4.10.   Particulate properties of mullite.

Composition of redox mixtures	Powder density (g cm^{-3})	Surface area (m^2 g^{-1})	Particle size[a] (nm)
30 g A + 1.6015 g silica fume + 12 g urea	2.750	45	48
10 g A + 0.5338 g silica fume + 5 g NH$_4$ClO$_4$ + 6.1 g urea	3.170	12.6	150

A = Al(NO$_3$)$_3$·9H$_2$O.
[a] From surface area.

The formation of mullite is confirmed by XRD, which shows the pattern of weakly crystalline mullite along with θ-Al$_2$O$_3$. This θ-Al$_2$O$_3$ phase disappears with heat-treatment. However, when extra oxidizer is incorporated the flame temperature is higher (>1400°C) and fully crystallized mullite is formed. Figure 4.15 shows the XRD pattern of as-prepared mullite using urea and extra oxidizer.

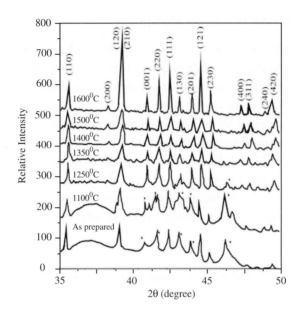

Fig. 4.15.   XRD patterns of heat treated mullite. Peaks with dot indicate θ-Al_2O_3.

Weakly crystalline mullite is seen in the as-prepared samples while complete mullitization occurs on heating at ~1300°C. The particulate properties like particle size and surface area are tabulated in Table 4.10.

The dilatometric studies of mullite compact show a total linear shrinkage of <3% between 1100°C and 1650°C with a shrinkage rate of ~25 μm/min. The sintering studies (1650°C/2 h) indicate lower density, probably due to the irregular plate morphology of the particles and also partly due to the appearance of flame during combustion, which induces mullite crystallization and neck growth processes. This powder when pressed uniaxially and sintered at 1600°C for 2 h achieves a theoretical density of 51%, while when cold isostatically pressed (210 MPa) and sintered at 1600°C achieves a density of 63%. Higher density (>95%) is achieved by hot isostatic pressing (100 MPa, 1500°C, 30 min) followed by sintering at 1700°C for 30 min.

Additives like MgO, ZrO_2, TiO_2, (0–5 wt.%) and rare earth oxides like Y_2O_3, La_2O_3, and CeO_2 are used to study their effect on the sintering behavior of mullite. With MgO, mullite shows a very dense microstructure with acicular grains (Fig. 4.16), while the microstructure of pure sintered mullite is composed of submicrometer equiaxed grains. It is observed that the one with additives shows elongated grains with large increase in grain size. Various grades of mullites are also prepared by heterogeneous solution combustion of redox mixtures containing different fuels and extra amounts of oxidizers and silica from a variety of sources. Sintering and microstructure of these mullites have been investigated in detail.[16]

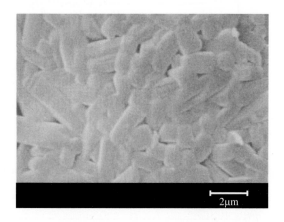

Fig. 4.16.   SEM of mullite (5 wt.% MgO) sintered at 1600°C.

4.10.2 *Al$_2$O$_3$ · SiO$_2$ System: Cordierite*

Cordierite is prepared by the combustion of an aqueous heterogeneous mixture containing magnesium nitrate/magnesium oxide (surface area 97 m^2 g^{-1}), aluminum nitrate, silica fume (surface area 200 m^2 g^{-1}), ammonium nitrate, and urea.[17] The mixture is stirred until the silica disperses fully. It is then placed in a cylindrical pyrex dish and introduced into a muffle furnace maintained at 500°C. The ignition and formation of the product occur in about 5 min.

The Al$_2$O$_3$·MgO formed *in situ* appears to react with fine-particle SiO$_2$ forming cordierite. The XRD patterns of as-prepared COR-I (from MgNO$_3$) and COR-II (from MgO) powders as well as heat treated (950–1300°C) COR-I and COR-II powders are shown in Figs. 4.17 and 4.18, respectively.

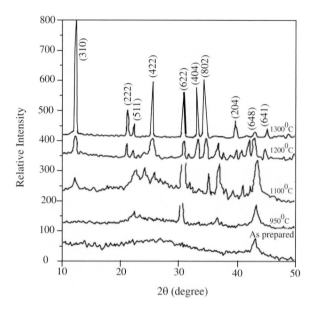

Fig. 4.17.   XRD patterns of as-prepared and heat-treated cordierite I.

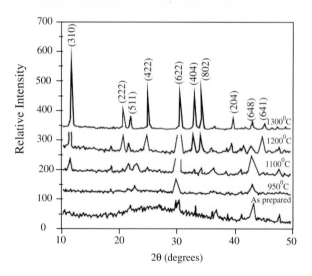

Fig. 4.18.   XRD patterns of as-prepared and heat-treated cordierite II.

The as-formed COR-I is essentially amorphous with a small amount of spinel, while the powders heated to 950°C show the presence of μ-cordierite in addition to spinel. Both the as-prepared and calcined COR-II powder show the presence of μ-cordierite and spinel. Powders of COR-I and COR-II, calcined at 1100°C, showed the presence of both μ- and α-cordierite but transformed completely to α-cordierite at 1250°C. Formation of μ-cordierite in the case of as-prepared COR-II is attributed to the presence of excess NH_4NO_3 in the redox mixture.

Particulate properties of the cordierite powders (COR-I and II) are summarized in Table 4.11.

Table 4.11.   Particulate properties of the cordierite powders.

Powder	Powder density $g\,cm^{-3}$	Foam density $g\,cm^{-3}$	Tap density $g\,cm^{-3}$	Surface area $m^2\,g^{-1}$	Particle size (nm)
COR-I	2.1	0.17	0.8	143	12
COR-II	2.2	0.24	1.2	44	81

Sintering properties and microstructure of combustion-derived cordierite powders are also investigated (Table 4.12). It is seen that magnesia-derived cordierite (COR-II) has better sintering characteristics compared to COR-I. A 97% theoretical density is attained by COR-II when sintered at 1450°C for 2 h as compared to 88% for COR-I when similarly sintered. This is also reflected in the microstructures of COR-I and COR-II (Figs. 4.19 and 4.20). COR-II shows the presence of macropores as it attains a higher bulk density at ~1440°C due to the liquid-phase sintering of low-melting eutectic in the $MgO–Al_2O_3–SiO_2$ system. The microhardness and thermal expansion coefficient of combustion-derived cordierite are comparable with those of sol–gel derived cordierite.

Table 4.12.   Properties of sintered cordierite compacts.

Powder	Green density[a] (% theoretical)	Bulk density[a] (% theoretical) Temperature (°C)			Linear shrinkage (% theoretical) Temperature (°C)					Coefficient of thermal expansion ($\alpha_{20-550}°C \times 10^6$, K^{-1})	Micro hardness (GPa)
		1425	1440	1450	1000	1245	1440	1450			
COR-I	58	84	86	88	14	15.6	15.8	16	—	—
COR-II	63	73	94	97	3.6	3.4	9.1	9.8	1.5	8.0

[a]Theoretical density of α-cordierite = 2.512 g cm^{-3}.

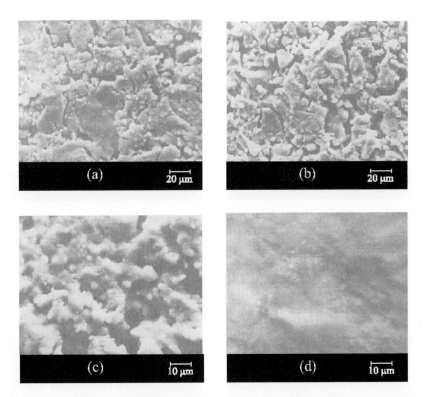

Fig. 4.19. SEM of cordierite surface (a) COR-I sintered at 1400°C, (b) COR-II sintered at 1400°C, (c) COR-I sintered at 1450°C, and (d) COR-II sintered at 1450°C.

4.10.3 *Al₂O₃:Si₃N₄ System: SiAlON*

The acronym "SiAlON" is given to the composite Al_2O_3:Si_3N_4, which is a solid solution of Al_2O_3 and Si_3N_4. The interest in SiAlON is due to its excellent thermochemical properties in corrosive and oxidative environments. In this composite, the Si and N atoms in the Si_3N_4 lattice get substituted by Al and O atoms, respectively, with the general formula $Si_{6-z}Al_zO_zN_{8-z}$ where the β-Si_3N_4 crystal structure is retained over the range $0 \leq Z \leq 4.2$.

SiAlON is prepared by the metathetical reaction of *in situ* formed alumina with α-Si_3N_4, and solid-state reaction between fumed alumina and

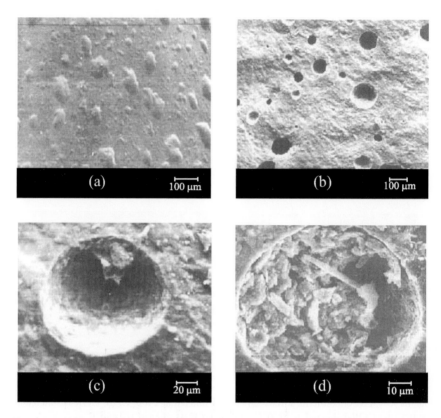

Fig. 4.20.   SEM of COR-II sintered at 1450°C: (a) surface showing drop like particles, (b) fractured surface showing distribution of macropores, (c) single almost empty macropore, and (d) partially filled macropore.

amorphous Si_3N_4.[18] The combustion reaction is carried out with α-Si_3N_4, $Al(NO_3)_3 \cdot 9H_2O$, and either urea or DFH as the fuel. The mixture when rapidly heated at 500°C ignites and burns with a flame temperature of 1000 ± 100°C. The composition of the redox mixture and the properties of SiAlON formed are summarized in Table 4.13.

The powder XRD pattern (Fig. 4.21) confirms single phase SiAlON formation and the grain morphology of the powder illustrates the fine particle nature (Fig. 4.22).

Table 4.13.   Composition of the redox mixtures and particulate properties of SiAlON.

Sample	Composition	Tap density $(g\,cm^{-3})$	Surface area $(m^2\,g^{-1})$	Particle size[a] (μm)
SiAlON-I	22.5 g A + 2.9 g α-Si$_3$N$_4$ + 9.0 g urea	0.58	39.8	2.4
SiAlON-II	22.5 g A + 2.9 g α-Si$_3$N$_4$ + 9.9 g DFH	0.40	46.2	1.6
SiAlON-III	5.67 g Fumed Al$_2$O$_3$ + 9.22 g α-Si$_3$N$_4$ + 2.37 g Al powder	0.34	50.47	1.2

A = Al(NO$_3$)$_3$·9H$_2$O.
[a] From surface area.

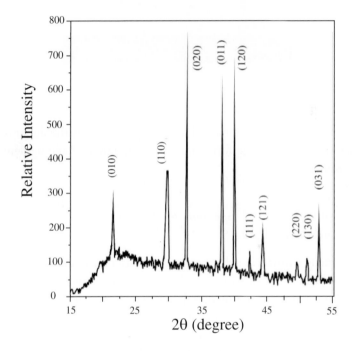

Fig. 4.21.   XRD pattern of β-SiAlON made by combustion method.

4.11 ALUMINA NANOCOMPOSITES

One of the important advantages of solution combustion synthesis is the ease of incorporating any desired amount of impurity atoms or ions in a given oxide matrix. This has been demonstrated by incorporating a variety

Fig. 4.22.   SEM of SiAlON.

of metal atoms/ions in alumina and aluminate matrix. As the dopants are of nanosize, they result in the formation of nanocatalysts, nanopigments, and nanophosphors etc.

4.11.1 *Nanocatalysts, Dispersion of Nano-metals (Ag, Au, Pd, and Pt) in Al$_2$O$_3$*

Nanosize metal particles as a distinct state of matter are of interest due to their catalytic activity. For a variety of catalytic reactions, highly dispersed Pt, Pd, Ag, and Au on alumina are used. Salts of Pt, Pd, Ag, and Au are known to decompose to their respective metal particles at relatively low temperatures (<500°C). Conventionally, fine alumina is soaked with solutions of these noble metal salts and heated for a long period to disperse metal oxides or metals. In case metal oxides are formed, the solid is then heated in hydrogen to convert the oxides into metal particles. This is a time consuming process as firstly alumina has to be made and then a series of steps involving metal salt impregnation, drying, heating, and hydrogen reduction are needed to disperse metal particles on alumina.

During the solution combustion synthesis of alumina with aluminum nitrate–urea mixture, the addition of 1–2% of noble metal salts in the redox mixture enable metal particles to disperse homogenously in the product. Due to the high exothermicity of the reaction, the noble metal halide and silver nitrate decompose to their respective metals. Using this concept, metal-dispersed alumina has been prepared and its catalytic property investigated.[19,20] The theoretical chemical reaction for the formation of metal-dispersed alumina may be represented by the following equation.

$$2Al(NO_3)_3(aq) + 5\underset{\text{Urea}}{CH_4N_2O(aq)}$$

$$\xrightarrow[\text{500°C}]{xH_2PtCl_6/PdCl_2/AgNO_3/HAuCl_4} M°/Al_2O_3(s) + 5CO_2(g)$$

$$+ 8N_2(g) + 10H_2O(g)$$

$$(\text{23 mol of gases/mol of } M°/Al_2O_3)$$

$$M° = Ag, Au, Pt, Pd \tag{22}$$

Combustion mixture for the preparation of 1% Pt/Al_2O_3 contains $Al(NO_3)_3$, H_2PtCl_6, and urea in the molar ratio of 1.98:0.02:4.95. In a typical preparation, 8.5 g of $Al(NO_3)_3 \cdot 9H_2O$, 0.112 g H_2PtCl_6, and 3.992 g urea are dissolved in 20 mL of water in a borosilicate dish of 130 cm³. The redox mixture is introduced into a muffle furnace preheated to 500°C. At the point of complete dehydration, the foam burns with a flame giving nano-Pt metal particle dispersed α-Al_2O_3. Similarly, Pd, Ag, and Au metal nanoparticles dispersed α-Al_2O_3 have also been prepared.

The dispersion of metal particles in alumina has been confirmed by powder XRD patterns of the products (Fig. 4.23). The particle size of metals in alumina has been determined by X-ray line broadening as well as TEM.

TEM of Pt and Pd metal particles dispersed over alumina by solution combustion method are given in Fig. 4.24. Uniform distribution of the metal particles is seen from the TEM pictures. A typical histogram of Pt particles and their size distribution curve determined from TEM is shown in Figs. 4.25(a) and 4.25(b), respectively. It is seen that the particles are in the range of 4–10 nm.

Due to their metal dispersion these materials (M^0/Al_2O_3 where M = Pt, Pd, Ag, Au) are good oxidation catalysts. Their three-way catalytic (TWC) activities, i.e., NO reduction, CO and CH_4 oxidation have been investigated by temperature-programmed reduction (TPR) studies.

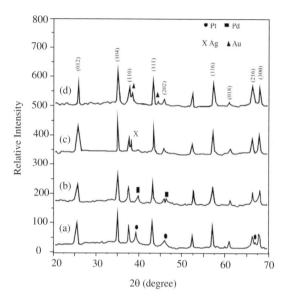

Fig. 4.23.   XRD patterns of: (a) 1% Pt/Al$_2$O$_3$, (b) 1% Pd/Al$_2$O$_3$, (c) 1% Ag/Al$_2$O$_3$, and (d) 1% Au/Al$_2$O$_3$.

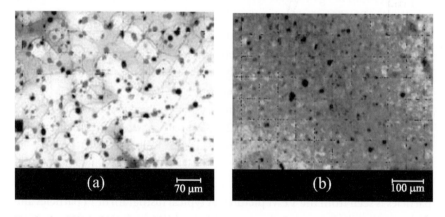

Fig. 4.24.   TEM of (a) 1% Pt/Al$_2$O$_3$ and (b) 1% Pd/Al$_2$O$_3$. Black dots indicate metal particles.

The results of these studies show that 100% CO conversion occurs below 300°C over supported Pt, Pd, Ag metals, whereas 90% conversion is observed over Au at 450°C (Fig. 4.26). Similarly, 100% NO conversion is seen over Pt and Pd below 400°C for NO + CO reaction, whereas ~90% NO is converted into N$_2$ above 650°C on Ag and Au (Fig. 4.27). Similarly, complete CH$_4$

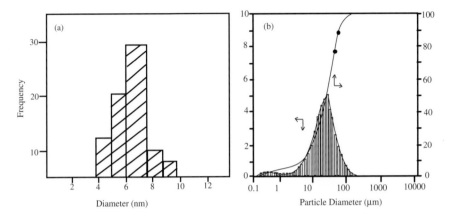

Fig. 4.25.   (a) Crystallite size distribution of Pt and (b) particle size distribution of Al_2O_3.

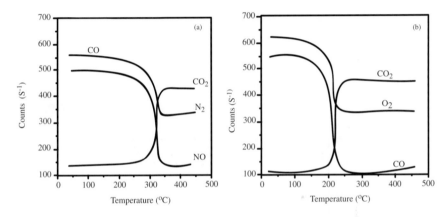

Fig. 4.26.   TPR curves of (a) CO oxidation profile with Pt/Al_2O_3 and (b) NO reduction by Pd/Al_2O_3.

oxidation occurs over Pd and Pt at 350°C and 425°C, respectively (Fig. 4.28), whereas complete oxidation of C_3H_8 over Pt and Pd gives CO_2 and H_2O at 175°C and 300°C, respectively.

The rate constants for CO oxidation are in the range of 8.6×10^3 and 2.5×10^3 cm^3 g^{-1} s^{-1} at 225°C for Pt/Al_2O_3 and Pd/Al_2O_3, respectively. The rate constants at 350°C for NO + CO reaction over these catalysts are 5.7×10^3 and 1.8×10^4 cm^3 g^{-1} s^{-1}, respectively.

For CH_4 oxidation the rate constants at 400°C and 410°C are 225 and 1700 cm^3 g^{-1} s^{-1} for Pt/Al_2O_3, and 210 and 530 cm^3 g^{-1} s^{-1} for Pd/Al_2O_3

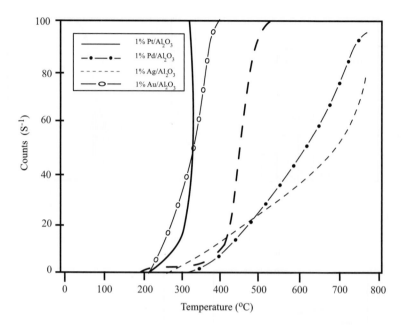

Fig. 4.27.   %NO conversion over M/Al$_2$O$_3$ (M = Pt, Pd, Ag, and Au).

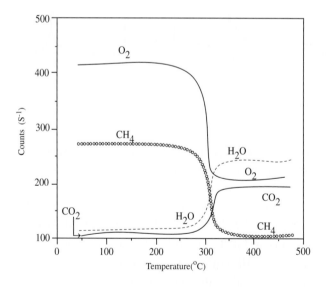

Fig. 4.28.   CH$_4$ oxidation over Pd/Al$_2$O$_3$.

respectively. The activation energies for $CO + O_2$ and $NO + CO$ reactions are in the range of 25–40 kJ mol^{-1} and that of CH_4 oxidation are 95.5 and 128.4 kJ mol^{-1} for Pt/Al_2O_3 and Pd/Al_2O_3, respectively.

Pt/Al_2O_3 and Pd/Al_2O_3 show better catalytic activities toward CO oxidation and NO reduction compared to Ag/Al_2O_3 and Au/Al_2O_3. The high rate constants and low activation energies arise due to large surface area and the presence of nanosize metal particles.

4.12 Nanopigments

Ceramic pigments are important due to their applications in tableware, sanitary ware, tiles, and glasses. They can be classified into two broad categories as idiochromatic and allochromatic pigments. Idiochromatic or self-colored pigments are those in which the transition metal ions are an essential part of the structure and contribute to the nature of the ligand field. Chrome green (Cr_2O_3) and Thenard's blue (cobalt aluminate: $CoAl_2O_4$) are some typical examples.

Allochromatic or other colored pigments derive their color from a foreign species and can be further categorized into substitution type pigments or inclusion type pigments. In substitution pigments, transition or lanthanide metal ions are present in the ligand field of the host lattice. The pink color of ruby ($Cr^{3+}/\alpha\text{-}Al_2O_3$), for example, is due to trace amounts of chromium present in the octahedral site of the corundum lattice. Inclusion pigments are those in which colored oxides are trapped or encapsulated in inert oxide hosts and can be considered as solid in solid type emulsion. For example, the well-known red zircon pigment ($Fe^{3+}/ZrSiO_4$) is red due to the entrapment of Fe^{3+} ion within a zircon crystal. It is interesting to note that the inclusions of nanoamounts of transition or rare earth compounds are enough to stain or color the entire oxides.

4.12.1 *Cobalt-Based Blue Alumina and Aluminates*

Both idiochromatic ($CoAl_2O_4$) and allochromatic ($Co_xAl_{2-x}O_4$) blue pigments are prepared by the solution combustion method. Idiochromatic cobalt aluminate (Thenard's blue) and nickel aluminate (sky blue) are prepared by the combustion of aqueous solutions containing stoichiometric quantities of corresponding metal nitrates and CH/ODH fuels.[21,22] Theoretical equations for the formation of the idiochromatic pigments can be written as follows:

$$M(NO_3)_2(aq) + 2Al(NO_3)_3(aq) + 5CH_6N_4O\,(aq)$$
$$\text{CH}$$
$$\xrightarrow{350°C} MAl_2O_4(s) + 15H_2O(g) + 5CO_2(g) + 14N_2(g)$$
$$(34\text{ mol of gases/mol of } MAl_2O_4)$$
$$M = Co, Ni \tag{23}$$

$CoAl_2O_4$ is indigo colored while nickel aluminate is light blue in color both with foamy textures. In case of $NiAl_2O_4$, the foam shows a fibrous structure and the product is voluminous having a foam density of $1.96 \times 10^{-3}\,g\,cm^{-3}$ and tap density of $0.02\,g\,cm^{-3}$. These features reflect the fluffy and fine nature of these pigments.

Allochromatic cobalt blue alumina and aluminates ($MgAl_2O_4$, $ZnAl_2O_4$, $CaAl_{12}O_{19}$, $SrAl_{12}O_{19}$) pigments are prepared by doping different atom% of Co^{2+} ions in these matrices. For example, 2.5 atom% Co-doped blue alumina is synthesized by rapidly heating aqueous solution containing 19.5 g $Al(NO_3)_3 \cdot 9H_2O$, 0.2 g $Co(NO_3)_2 \cdot 6H_2O$ and 8 g urea at 500°C. The solution burns with a flame yielding 2.8 g of foamy cobalt-doped blue alumina. Chemical equations for the formation of allochromatic blue pigments $Co_xAl_{2-x}O_3$ assuming complete combustion is written as

$$2Al(NO_3)_3(aq) + 5CH_4N_2O(aq)$$
$$\text{Urea}$$
$$\xrightarrow[350\pm10°C]{Co(NO_3)_2} Co_xAl_{2-x}O_3(s) + 5CO_2(g) + 10H_2O(g) + 8N_2(g)$$
$$(23\text{ mol of gases/mol of } Co_xAl_{2-x}O_3) \tag{24}$$

Doping different atom% ($x = 0.5, 2.5, 7.5, 10$) of Co produces various shades of blue alumina. However, small amount of Co ion in +2 state with four coordination crystallizes in the matrix of α-Al_2O_3 to give a uniform blue color (Fig. 4.29). The X-ray pattern of 0.5 atom% Co^{2+} in Al_2O_3 shows a single phase of α-Al_2O_3. But with increase of Co ion at 5 atom% and beyond, the $CoAl_2O_4$ phase starts appearing becoming distinct at 10 atom%. For Co^{2+} substitution of $x = 0.01$–0.1, lattice parameter a value of α-Al_2O_3 structure changes from 0.4772 to 0.4767 nm and c values changes from 1.3043 to 1.3013 nm. The increase in Co content causes a decrease in the lattice

Fig. 4.29. Co^{2+} doped Al_2O_3 (blue pigment).

constants, indicating an increase in the crystal field around the Co^{2+} ion in the host lattice.

Similarly, Co^{2+}-doped in $MgAl_2O_4$, $ZnAl_2O_4$, and $CaAl_{12}O_{19}$, blue pigments are prepared by solution combustion of corresponding metal nitrates and urea mixtures. Chemical equation for the formation of allochromatic blue pigments $Co_xM_{1-x}Al_2O_4$ (M = Mg and Zn, $x = 0.01–0.2$) assuming complete combustion is written as

$$3M(NO_3)_2(aq) + 6\,Al(NO_3)_3(aq) + 20CH_4N_2O\,(aq)$$
$$\underset{Urea}{}$$

$$\underset{500°C}{\overset{Co(NO_3)_2}{\rightleftharpoons}} 3Co_xM_{1-x}Al_2O_4(s) + 40H_2O(g) + 20CO_2(g) + 32N_2(g)$$

$$(\sim31\text{ mol of gases/mol of }Co_xM_{1-x}Al_2O_4) \quad (25)$$

Figure 4.30 shows the voluminous blue colored pigment with 1 atom% of cobalt in $MgAl_2O_4$. In the same way Co-doped calcium hexaaluminate ($CaAl_{12}O_{19}$) leads to the production of a blue pigment (Fig. 4.31).

The particulate properties of blue alumina pigments are summarized in Table 4.14.

The diffuse reflectance spectrum of $Co_xAl_{2-x}O_3$ and $Co_xM_{1-x}Al_2O_4$ is characterized by a broad absorption band whose centroid is placed in the red part of the visible spectrum (Fig. 4.32). The intense convoluted absorption band with prominent absorbance peaks at around 550, 580, and 620 nm indicates tetrahedral co-ordination of Co^{2+}. The wide and intense absorption band between 500 and 700 nm could be attributed to the γ_3 transition from $^4A_2$ (F) ground state to the excited $^4T_1$ (P) state.

Fig. 4.30.   One percent Co^{2+} in $MgAl_2O_4$.

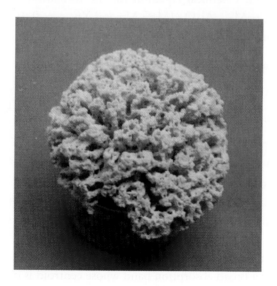

Fig. 4.31.   One percent Co^{2+}: $CaAl_{12}O_{19}$.

Table 4.14.   Particulate properties of blue alumina pigments.

Pigments (1 atom% of dopant)	Powder density $(g\,cm^{-3})$	Surface area $(m^2\,g^{-1})$	Particle size[a] (nm)
$Co^{2+}: \alpha\text{-}Al_2O_3$	2.75	10	220
$Co^{2+}: MgAl_2O_4$	2.32	20	100
$Co^{2+}: ZnAl_2O_4$	2.70	12	140
$Co^{2+}: CaAl_{12}O_{19}$	2.35	8	260
$CoAl_2O_4$	2.20	58	30

[a] From surface area.

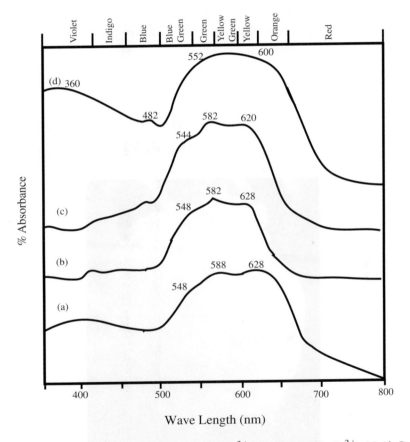

Fig. 4.32.   Diffused reflectance spectra of: (a) $Co^{2+}: ZnAl_2O_4$, (b) $Co^{2+}: MgAl_2O_4$, (c) $Co^{2+}: \alpha\text{-}Al_2O_3$, and (d) $CoAl_2O_4$.

4.12.2 *Chromium-Doped Pink Alumina (Cr^{3+}/Al_2O_3): Ruby*

Conventionally, the formation of ruby, chromium-doped α-alumina (Cr^{3+}/Al_2O_3) is known to take place only when heated above 1350°C for 96 h.[23] However, it can be prepared within 5 min by the solution combustion method.[24] Ruby powder (Cr^{3+}-doped Al_2O_3) is prepared by the combustion of an aqueous redox mixture containing stoichiometric amounts of aluminum nitrate, chromium nitrate, and urea. In a typical experiment, 20 g $Al(NO_3)_3 \cdot 9H_2O$, 0.0148 g $Cr(NO_3)_3 \cdot 6H_2O$, and 8 g of urea are dissolved in 15 mL of water in a Pyrex dish which on combustion burn with an incandescence flame and form voluminous pink foamy alumina powder (Fig. 4.33). Ruby is also prepared by using other fuels like DFH, HMT, and ODH.[4,5,7]

Assuming complete combustion, the theoretical equation for the formation of pink alumina may be written as follows:

$$2Al(NO_3)_3(aq) + 5\underset{\text{Urea}}{CH_4N_2O}(aq)$$

$$\xrightarrow[350\pm10°C]{Cr(NO_3)_3} Cr_xAl_{2-x}O_3(s) + 5CO_2(g) + 10H_2O(g) + 8N_2(g)$$

$$\text{(23 mol of gases/mol of ruby)} \qquad (26)$$

Fig. 4.33.   One atom% Cr in α-Al_2O_3: ruby.

Both Cr_2O_3 and Al_2O_3 crystallize in the same corundum structure and in principle, they can form solid solution in the entire range. However, for the synthesis of ruby, less than 1% Cr in Al_2O_3 is sufficient to give an intense pink color. It is possible to dope up to 2 atom% of Cr by solution combustion method due to the short duration of the combustion reaction (less than 2 min). The solution combustion process facilitates the homogeneous atomic level substitution of chromium atoms.

4.12.3 *Chromium-Doped Aluminates and Orthoaluminates* $(Cr^{3+}/MAl_2O_4(M = Mg \ \& \ Zn) \ and \ LaAlO_3)$

Pink colored Cr^{3+}-doped $MgAl_2O_4$ is prepared by the combustion of an aqueous redox mixture containing 6.83 g $Mg(NO_3)_2 \cdot 6H_2O$, 0.007 g $Cr(NO_3)_3 \cdot 6H_2O$, and 10.7 g Urea at 500°C in less than 5 min. Assuming complete combustion, the theoretical equation for the formation of chromium-doped pink aluminate may be written as follows:

$$M(NO_3)_2(aq) + 2Al(NO_3)_3(aq) + \underset{\text{Urea}}{10CH_4N_2O(aq)}$$

$$\xrightarrow[350 \pm 10°C]{Cr(NO_3)_3} Cr_xMAl_{2-x}O_4(s) + 10CO_2(g) + 20H_2O(g) + 14N_2(g)$$

$$\text{(44 mol of gases/mol of pink aluminate)} \quad (27)$$

Similarly, pink colored chromium-doped $LaAlO_3$ is prepared by solution combustion method using a redox mixture containing 23.08 g $La(NO_3)_3 \cdot 6H_2O$, 0.007 g $Cr(NO_3)_3 \cdot 6H_2O$, and 16 g urea/14.4 g CH fuel.[24]

The as-prepared powders are X-ray crystalline. Typical XRD pattern of $Cr^{3+}/MgAl_2O_4$ is shown in Fig. 4.34. The lattice parameters $a = 0.811$ nm ($MgAl_2O_4$) calculated from the XRD pattern are in good agreement with the values quoted in literature ($a = 0.808$ nm).

The particulate properties of pink alumina pigments are summarized in Table 4.15

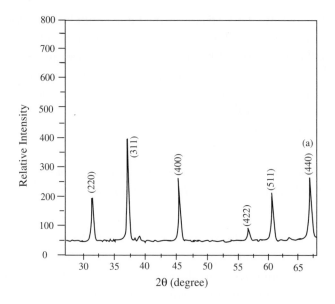

Fig. 4.34.   XRD pattern of flourescent $Cr^{3+}/MgAl_2O_4$ (Cu K_α radiation).

Table 4.15.   Particulate properties of pink alumina pigments.

Pigments (1 atom% of dopant)	Powder density $(g\,cm^{-3})$	Surface area $(m^2\,g^{-1})$	Particle size[a] (nm)
Cr^{3+}: α-Al_2O_3	2.80	9	240
Cr^{3+}: $MgAl_2O_4$	3.15	21	20
Cr^{3+}: $ZnAl_2O_4$	3.52	11	160
Cr^{3+}: $LaAlO_3$	4.89	4	350

[a] From surface area.

4.13 NANOPHOSPHORS

Luminescence is the emission of light from sources other than thermal energy, involving bioluminescence, photoluminescence, cathode luminescence, etc. It includes both phosphorescence and fluorescence. Fluorescence is the emission of visible light from materials irradiated usually from higher frequency source or from impact of electrons as in phosphors. Phosphor is a generic name for the class of substances which exhibit luminescence.[25]

Inorganic luminescence materials are crystalline compounds that absorb energy and subsequently emit this absorbed energy as light. Phosphors are composed of an inert host lattice and an optically excited activator, typically a 3d or 4f electron metal such as Ce^{3+}, Cr^{3+}, Eu^{3+}, and Tb^{3+}. Oxide phosphors are found to be suitable for information display devices such as field emission display (FED), vacuum fluorescent display (VFD), electroluminescent (EL) devices, plasma panel display (PDP) devices, TV receivers, cathode ray tube (CRT), fluorescent lamps, scintillation counters, etc.

Recently, nanoscale phosphors like ruby (Cr^{3+}:Al_2O_3), Cr^{3+}/YAG have been investigated for the measurement of temperature as they exhibit linear response with temperature. The technique relies on the intensity and peak position change of luminescent crystalline materials. The nonradiative transition rate and lifetime of the excited state correspond to the temperature, and can be used in a noncontact thermometer with the support of a fiber-optic sensor. Such a sensor is composed of a light source for a crystal excitation and a fluorescence intensity or decay time detector to determine temperature.[26] This is a powerful tool for temperature measurement.

The preparation of chromium-doped alumina (Cr^{3+}/Al_2O_3 — ruby), aluminates, and orthoaluminates (Cr^{3+}/$MgAl_2O_4$ and $LaAlO_3$) were discussed in the previous sections (Sec. 4.12.1 and 4.12.2). The presence of chromium ions in the host lattice is confirmed by their fluorescence spectra and characteristic color. The combustion synthesized 0.05% Cr^{3+}/Al_2O_3, $MgAl_2O_4$, and $LaAlO_3$ show emission bands at 695, 687 and 734 nm, respectively. Also, the excitation bands at 406, 548 nm (ruby); 395, 584 nm (spinel); and 514 nm (orthoaluminate) agree well with the literature values (Fig. 4.35). Although all these lattices have Cr^{3+} ions in octahedral co-ordination, they show emission at different frequencies. This could be attributed to the difference in lattice pressures experienced by the Cr^{3+} ions.

Fluorescence decay times measured using pulsed nitrogen laser (150 KW, 8 ns, and 10 pps) are as follows: Cr^{3+}/Al_2O_3(3.6 ms), Cr^{3+}/$MgAl_2O_4$ (3.9 ms), and Cr^{3+}/$LaAlO_3$ (52.0 ms). Cr^{3+}/$LaAlO_3$ is of particular interest due to its longer decay time. Similarly chromium-doped YAG (Cr^{3+}:$Y_3Al_5O_{12}$) is prepared and investigated as a red-emitting phosphor. The effects of processing parameters like the fuels used, sintering temperature, particle size, etc., on the luminescent properties of the phosphor have also been examined.[27]

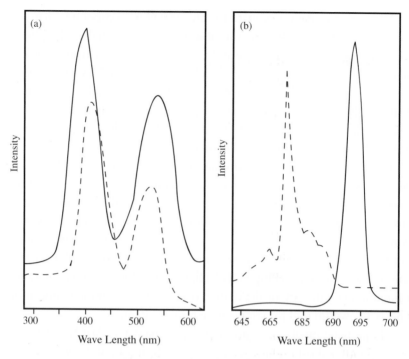

Fig. 4.35.   Fluorescence spectra of Cr^{3+}/Al_2O_3 (—) and $Cr^{3+}/MgAl_2O_4$ (---) (a) excitation spectrum and (b) emission spectrum.

4.13.1 *Phosphor Materials (Luminescence in Aluminum Oxide Hosts)*

Fluorescent lamps are a typical example of phosphors being coated on glass substrate using suspension of phosphor powder particles. Production of good phosphors requires rigorous attention during synthesis or activation process. Careful control of activator concentrations, usually measured in parts per million is essential. Multiple activators may be necessary to achieve the desired color, brightness, response, or decay times. Conventionally, these materials are made using solid-state method which requires high-temperature ($>1000°C$) processing. For example, $CeMgAl_{11}O_{19}$ a green phosphor is prepared from a mixture of Al_2O_3, $MgCO_3$, CeO_2, and Tb_4O_7 heated at $1500°C$, for 5 h with small quantities of MgF_2 or AlF_3 as mineralizer. This method however, has several limitations such as

(i) Inhomogeneity in the product formed.
(ii) Formation of large particles with low surface area which needs mechanical particle size reduction introducing impurity and defects.
(iii) Presence of defects, which interfere with luminescence.

However, these problems can be overcome by using wet chemical methods which are much more involved and cumbersome, requiring expensive starting materials like alkoxides.

The solution combustion process is used for the preparation of both phosphors and luminescent oxide materials by atomic level doping of various rare earth ions like Ce^{3+}, Tb^{3+}, Eu^{2+} and chromium in different aluminum oxide hosts such as barium and magnesium hexaaluminates. Solution combustion method being highly exothermic facilitates the incorporation of the desired activators in the aluminum oxide matrix.

Blue Phosphors: The activator used for blue phosphor is Eu^{2+} and the commonly used hosts are $BaMgAl_{10}O_{17}$, $BaMg_2Al_{16}O_{27}$, and $Ba_{0.64}Al_{12}O_{18.64}$.[28] All these phosphors are prepared by the solution combustion process and their luminescent properties investigated. An aqueous concentrated solution containing stoichiometric amounts of metal nitrates $M(NO_3)_2$ (M=Ba and Mg); $M(NO_3)_3$ (M=Al and Eu) and a fuel like CH or DFH or urea undergo combustion reaction at 400–500°C to give the required blue phosphor. Assuming complete combustion, the theoretical equations for the formation of $Ba_{0.64-x}Eu_xAl_{12}O_{18.64}$ and $Ba_{1-x}Eu_xMgAl_{10}O_{17}$ can be written as follows:

$$(0.64 - x)Ba(NO_3)_2 + xEu(NO_3)_3 + 12Al(NO_3)_3 + \underset{CH}{23.3CH_6N_4O}$$

$$\longrightarrow \underset{\text{Blue phosphor}}{Ba_{0.64-x}Eu_xAl_{12}O_{18.64}(s)} + pCO_2(g) + qN_2(g) + rH_2O(g) \quad (28)$$

$$(1 - x)Ba(NO_3)_2 + xEu(NO_3)_3 + Mg(NO_3)_2$$
$$+ 10Al(NO_3)_3 + \underset{Urea}{28.33CH_4N_2O}$$

$$\longrightarrow \underset{\text{Blue phosphor}}{Ba_{1-x}Eu_xMgAl_{10}O_{17}(s)} + pCO_2(g) + qN_2(g) + rH_2O(g)$$

$$(29)$$

The powder as such is colorless, and on exposure to UV radiation shows blue color (Fig. 4.36). The powder XRD patterns confirmed the formation of blue phosphors ($Ba_{0.64-x}Eu_xAl_{12}O_{18.64}$; x=0.64–1.3) (Fig. 4.37). It is observed that hexaaluminate having barium content ≤ 0.82 does not show any additional XRD peaks corresponding to $BaAl_2O_4$. When $x > 0.82$, additional peaks corresponding to those of $BaAl_2O_4$ are noticed. No splitting of (107) reflection is observed for the as-prepared and calcined phosphors indicating that the phosphors obtained by combustion process are mono-phasic nonstoichiometric hexaaluminate.

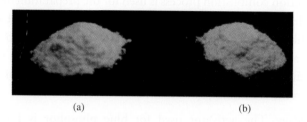

(a)   (b)

Fig. 4.36. Photograph of as-prepared (a) $BaMgAl_{11}O_{17}$: Eu^{2+} and (b) $BaMgAl_{10}O_{17}$: Mn^{2+} under UV irradiation.

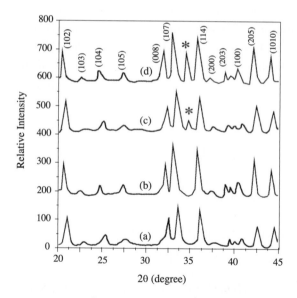

Fig. 4.37. XRD patterns of Eu^{2+}-doped $xBaO \cdot 6Al_2O_4$: (a) $x = 0.64$; (b) $x = 0.82$; (c) $x = 1.00$, and (d) $x = 1.3$, where $^*BaAl_2O_4$.

The emission spectra of as-prepared Eu^{2+}-activated $BaMgAl_{10}O_{17}$, $BaMg_2Al_{16}O_{27}$, and $Ba_{0.64}Al_{12}O_{18.64}$ show characteristic blue emission at 450, 450, and 435 nm, respectively (Fig. 4.38). The optimum concentration of activator for maximum emission intensity is found to be 16 mol%. This emission band is attributed to $4f^65d \rightarrow 4f^7$ transition of Eu^{2+}. With increasing barium content the emission band shifts toward longer wavelength because of the formation of Eu^{2+}-doped barium aluminate.

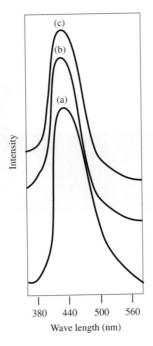

Fig. 4.38.   Emission spectra of Eu^{2+}-activated (a) $BaMgAl_{10}O_{17}$, (b) $BaMg_2Al_{16}O_{27}$, and (c) $Ba_{0.64}Al_{12}O_{18.64}$ blue phosphors (urea process). Excitation wave length was 254 nm.

Green Phosphors: The most commonly used activator for green emission is Tb^{3+}. The oxide host employed for Tb^{3+} green emission is $(La,Ce)MgAl_{11}O_{19}$. The other green phosphors are $(Ce,Gd)MgB_5O_{10}:Tb^{3+}$ and $(La,Ce)PO_4:Tb^{3+}$. Tb^{3+} and Mn^{2+} cannot be excited directly as the charge transfer band of Tb^{3+} and Mn^{2+} lie well above the predominant mercury line (254 nm). Hence, it is necessary to use suitable sensitizers like Ce^{3+} and Eu^{2+} involving intense absorption in the 254 nm region.

Tb^{3+} activated (La,Ce)MgAl$_{11}$O$_{19}$ green phosphor is obtained by rapidly heating an aqueous concentrated solution containing stoichiometric amounts of metal nitrates of La, Ce, Tb, Mg, and urea/DFH redox mixture at 500°C (Fig. 4.39). Redox compositions of M(NO$_3$)$_2$:urea (1:1.66), M(NO$_3$)$_3$:urea (1:2.5), and M(NO$_3$)$_2$:DFH (1:1.88) are used for combustion.[29] Theoretical equation assuming complete combustion can be written as follows:

$$(1 - x)Ce(NO_3)_3 + xTb(NO_3)_3 + Mg(NO_3)_2$$
$$+ 11Al(NO_3)_3 + 31.67\underset{\text{Urea}}{CH_4N_2O}$$
$$\longrightarrow \underset{\text{Green phosphor}}{Ce_{1-x}Tb_xMgAl_{11}O_{19}} + pCO_2(g) + qN_2(g) + rH_2O(g)$$

$$(30)$$

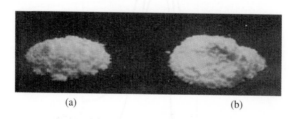

(a)   (b)

Fig. 4.39. Photograph of as-prepared (a) LaMgAl$_{11}$O$_{19}$: Ce^{3+}, Tb^{3+} and (b) BaMgAl$_{10}$O$_{17}$:Eu^{2+}, Mn^{2+} under UV irradiation.

The powder XRD pattern of (La$_{0.98}$Ce$_{0.5}$Tb$_{0.35}$)MgAl$_{11}$O$_{19}$ in Fig. 4.40 reveals single-phase crystalline nature. The emission spectrum of combustion-synthesized (La$_{0.98}$Ce$_{0.02}$)MgAl$_{11}$O$_{19}$ shows a broad band at 340 nm (Fig. 4.41). The emission of Ce^{3+} is due to 4f^{6}5d $\rightarrow$ ^{2}F$_j$ (with $j = 5/2$ and 7/2) transition. Addition of Tb^{3+} in (La$_{0.98}$Ce$_{0.02}$)MgAl$_{11}$O$_{19}$ results in emissions at 480 and 543 nm in addition to Ce^{3+} emission. The Tb^{3+} emissions, which arise due to energy transfer from Ce^{3+} are attributed to ^{5}D$_3$ $\rightarrow$^{7}F$_5$ and ^{5}D$_3$ $\rightarrow$^{7}F$_5$ transitions.

The particulate properties of solution combustion-derived lamp phosphors are listed in Table 4.16.

The rare earth doped phosphors satisfy all the essential requirements of lamp phosphors such as fine and ultrafine particle size ($<3\,\mu$m), large surface area, homogeneous distribution of the rare earth ion in the host lattice, etc.

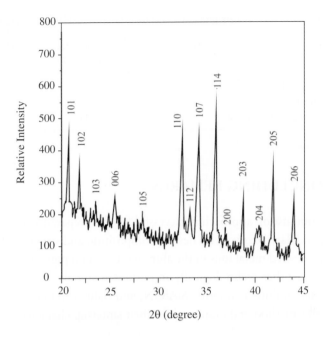

Fig. 4.40.   XRD pattern of $(La_{0.15}Ce_{0.5}Tb_{0.35})MgAl_{11}O_{19}$ (urea process).

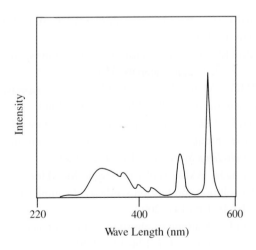

Fig. 4.41.   Fluorescence spectra of $LaMgAl_{11}O_{19}$:Ce^{3+},Tb^{3+}.

Table 4.16.   Particulate properties of combustion-derived lamp phosphors.

Compound	Powder density $(g\,cm^{-3})$	Surface area $(m^2\,g^{-1})$	Particle size[a] (μm)
$Ba_{0.64}Al_{12}O_{18.64}{:}Eu^{2+}$ (CH)	2.28	22	0.119
$BaMgAl_{10}O_{17}{:}Eu^{2+}$ (urea)	3.60	10	0.166
$Ce_{0.67}Tb_{0.33}MgAl_{11}O_{19}$ (urea)	4.10	11	0.133

[a] From surface area.

4.14 CONCLUDING REMARKS

Solution combustion approach is successful for the preparation of a variety of alumina and related oxides with desired composition and structure. It is possible to dope desired metal ions in the aluminas to obtain pigments, phosphors, catalysts, and other technologically important materials. High-temperature materials like mullite, cordierite, SiAlON, and tialite ceramics are easily prepared by this method and examined for their sintering characteristics.

References

1. Gitzen WH (ed.), *Alumina as a Ceramic Material*, The American Ceramic Society, Inc., Columbus, OH, 1970.
2. Kingsley JJ, Patil KC, A novel combustion process for the synthesis of fine particle α-alumina and related oxide materials, *Mater Lett* **6**: 427–432, 1988.
3. Kiminami RHGA, Morelli R, Folz DC, Clark DE, Microwave synthesis of alumina powder, *Am Ceram Soc Bull* March, 63–67, 2000.
4. Chandran RG, Patil KC, Chandrappa GT, Combustion synthesis of oxide materials using metal nitrates — Diformyl hydrazine redox mixtures, *Int J Self-Propagating High-Temp Synth* **3**(2): 131–142, 1994.
5. Prakash AS, Khadar AMA, Patil KC, Hegde MS, Hexamethylene tetramine: A new fuel for solution combustion synthesis of complex metal oxides, *J Mater Synth Process* **10**(3): 135–141, 2002.
6. Mimani T, Patil KC, Solution combustion synthesis of nanoscale oxides and their composites, *Mater Phys Mech* **4**: 134–137, 2001.
7. Suresh K, Patil KC, A recipe for an instant synthesis of fine particle oxide materials, in: Rao KJ (ed.), *Perspectives in Solid State Chemistry*, Norosa Pub. House, New Delhi, pp. 376–388, 1995.

8. Arul Dhas N, Patil KC, Combustion synthesis and properties of zirconia–alumina powders, *Ceram Int* **20**: 57–66, 1994.

9. Kingsley JJ, Suresh K, Patil KC, Combustion synthesis of fine-particle metal aluminates, *J Mater Sci* **25**: 1305–1312, 1990.

10. Mimani T, Instant synthesis of nanoscale spinel aluminates, *J Alloys Compd* **315**: 123–128, 2001.

11. Kingsley JJ, Suresh K, Patil KC, Combustion synthesis of fine particle rare earth orthoaluminates and yttrium aluminum garnet, *J Solid State Chem* **87**: 435–442, 1990.

12. Ekambaram S, Arul Dhas N, Patil KC, Synthesis and properties of aluminum borate (a light weight ceramic), *Intl J Self Propagating High-Temp Synth* **4**: 85–93, 1995.

13. Sekar MMA, Patil KC, Synthesis and properties of tialite, β-Al_2TiO_5, *Br Ceram Trans* **93**: 146–149, 1994.

14. Arul Dhas N, Patil KC, Synthesis of $AlPO_4$, $LaPO_4$ and $KTiOPO_4$ by flash combustion, *J Alloys Compd* **202**: 137–141, 1993.

15. Gopi Chandran R, Patil KC, A rapid combustion process for the preparation of crystalline mullite, *Mater Lett* **10**: 291–295, 1990.

16. Gopi Chandran R, Chandrashekar BK, Ganguly C, Patil KC, Sintering and microstructural investigations on combustion processed mullite, *J Eur Ceram Soc* **16**: 843–849, 1996.

17. Gopi Chandran R, Patil KC, Combustion synthesis, characterization, sintering and microstructure of cordierite, *Br Ceram Trans* **92**: 239–245, 1993.

18. Rajan TSK, Murthy HSGK, Patil KC, Combustion synthesis and properties of β-SiAlON, *6th Annual general body meeting of Materials Research Society of India, IIT Kharagpur*, 1985.

19. Bera P, Patil KC, Jayram V, Hegde MS, Subbanna GN, Combustion synthesis of nano-metal particles supported on α-Al_2O_3: CO oxidation and NO reduction catalysts, *J Mater Chem* **9**: 1081–1085, 1999.

20. Bera P, Patil KC, Hegde MS, Oxidation of CH_4 and C_3H_8 over combustion synthesized nanosize metal particles supported on α-Al_2O_3, *Phys Chem Chem Phys* **2**: 373–378, 2000.

21. Mimani T, Samrat Ghosh, Combustion synthesis of cobalt pigments: Blue and pink, *Curr Sci* **7**: 892–896, 2000.

22. Patil KC, Ghosh S, Aruna ST, Ekambaram S, Ceramic pigments: Solution combustion approach, *The Indian Potter* **34**: 1–9, 1996.

23. De Biasi RS, Rodrigues DCS, Inhomogeneous distribution of Cr impurities in α-Al_2O_3 during refractory aging, *J Mater Sci* **16**: 968–972, 1981.

24. Kingsley JJ, Manickam N, Patil KC, Combustion synthesis and properties of fine particle fluorescent aluminous oxides, *Bull Mater Sci* **13**: 179–189, 1990.

25. Blasse G, Grabmair BC, *Luminescent Materials*, Springer Verlag, Berlin, 1994.

26. Lee J, Kotov NA, Thermometer design at the nanoscale, *Nanotoday* **2**: 48–51, 2007.

27. Shea LE, McKittrick J, Lopez OA, Synthesis of red-emitting, small particle size luminescent oxides using an optimized combustion process, *J Am Ceram Soc* **79**: 3257–3265, 1996.

28. Ekambaram S, Patil KC, Maaza M, Synthesis of lamp phosphors: Facile combustion approach, *J Alloys Compd* **393**: 81–92, 2005.

29. Ekambaram S, Patil KC, Combustion synthesis and properties of Tb^{3+} and Mn^{2+}-activated green phosphors, *J Mater Synth Process* **4**: 334–337, 1996.

Chapter 5

Nano-Ceria and Metal-Ion-Substituted Ceria

5.1 INTRODUCTION

Ceria (CeO_2) is a pale yellow-white powder with a fluorite structure and lattice parameter $a = 0.5411$ nm possessing a high density of $7.215\,g\,cm^{-3}$. The fluorite structure can be described in terms of a cubic close packing of large cations in which all the tetrahedral voids are filled by the anions (Fig. 5.1). As a result a cubic eightfold coordination of the cations is produced. Bonding extends uniformly in all three dimensions leading to a strong covalent character. Accordingly, fluorite type oxides such as CeO_2, ZrO_2, and ThO_2 tend to be refractory and have high melting points of about 2100°C.

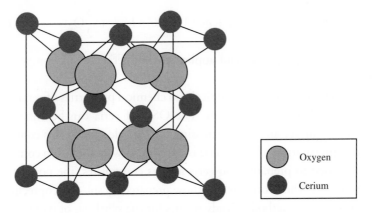

Fig. 5.1.   Fluorite structure of CeO_2.

Currently, ceria is attracting a great deal of attention due to its use as a "three-way catalyst" (TWC) converter (CO, hydrocarbons oxidations, NO_x

reduction) in automobiles. Some important features of ceria that makes it an effective catalyst are

(i) Oxygen storage capacity (OSC), due to a facile Ce^{+4}/Ce^{+3} redox reaction.
(ii) Ability to improve the dispersion of the noble metal.
(iii) Thermal stabilization of the alumina support.
(iv) Promotion of the water gas shift reaction.
(v) Direct interaction with the noble metal, which leads to an improved CO and hydrocarbon oxidation.

Among the above-mentioned features, the OSC is particularly relevant. It may be noted that TWCs usually achieve simultaneous conversions of the three main pollutants in auto exhausts in a rather narrow range of the air-to-fuel ratio, centered at its stoichiometric value ($\sim$14.6). While in real car operations the air-to-fuel ratio can widely oscillate around the above optimal value even with electronic controls in place. Therefore, one of the main goals of the catalytic converter is of enlarging the so called "operational window." The redox reaction (Eq. 1) is a measure of air-to-fuel ratio, which regulates the consumption of excess oxygen under "lean" mixture and releasing it under "rich" mixture conditions.[1,2]

$$CeO_2 + \delta CO \longrightarrow CeO_{2-\delta} + \delta CO_2; \quad CeO_{2-\delta} + \frac{\delta}{2}O_2 \longrightarrow CeO_2 \quad (1)$$

Some important properties and applications of ceria and its composites are summarized in Table 5.1

It is well known that both physical and chemical properties of ceria depend on the method of its preparation. The OSC of ceria can be enhanced by increasing its surface area and decreasing the particle size. In this context, the method of preparation of nano-ceria is important. When solution combustion method is used to prepare nano-ceria and its composites, the powders show better reactivity when compared to those prepared by sol–gel, coprecipitation and hydrothermal methods. Ready recipes for the synthesis of nano-ceria and metal-substituted ceria by the solution combustion method are enumerated in this chapter. The particulate nature, structure, morphology, and catalytic properties of combustion-derived nano-ceria and its composites are presented. The performance of metal-substituted ceria as a three-way automobile exhaust

Table 5.1. Properties and applications of ceria and its composites.

Material	Property	Applications
Ceria	Redox property — OSC	Oxidation–reduction catalysts
Metal dispersed ceria (Pt, Pd, Rh, Ru, etc.)	Redox property — OSC	Oxidation–reduction catalysts
Doped ceria (Gd_2O_3–CeO_2)	Ionic conductivity	Electrolyte for SOFCs
Nano-ceria	Abrasive	Polishing lenses, gems, TV tubes, etc.
	Optical property	Sunscreens, coatings for corrosion protection and self-cleaning ovens at high temperatures
Metal substituted nano-ceria ($Ce_{1-x}M_xO_{2-\delta}$)	Redox property — OSC	Catalysts
Solid solutions of ceria: CeO_2–ZrO_2	Mechanical — OSC	Stabilization of c-ZrO_2
Solid solutions of ceria: CeO_2–ThO_2	Thermal radiation	High-temperature gas mantles

catalyst has been demonstrated by coating cordierite honeycomb monolith in a single step.

5.2 SYNTHESIS AND PROPERTIES OF NANO-CERIA

Nano-ceria is prepared by the combustion of aqueous solution containing stoichiometric amounts of ceric ammonium nitrate with either urea (U)/carbohydrazide (CH)/oxalyl dihydrazide (ODH)/tetraformal trisazine (TFTA)/glycine (GLY) redox mixture. The solution undergoes flaming combustion to give nano-ceria (Fig. 5.2).[3,4]

Assuming complete combustion, the theoretical equations for the formation of ceria using ceric ammonium nitrate and ODH/GLY are written as follows:

$$5(NH_4)_2Ce(NO_3)_6(aq) + 12\underset{(ODH)}{C_2H_6N_4O_2}(aq)$$

$$\longrightarrow 5CeO_2(s) + 44N_2(g) + 24CO_2(g) + 56H_2O(g)$$

$$(24.8 \text{ mol of gases/mol of } CeO_2) \qquad (2)$$

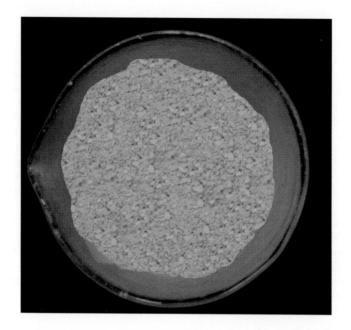

Fig. 5.2.   As-formed CeO_2 from combustion of redox mixture.

$$(NH_4)_2Ce(NO_3)_6(aq) + 4C_2H_5NO_2(aq) + 6NH_4NO_3(aq)$$
$$\text{(GLY)}$$
$$\longrightarrow CeO_2(s) + 8CO_2(g) + 26H_2O(g) + 12N_2(g)$$
$$(46 \text{ mol of gases/mol of } CeO_2) \tag{3}$$

Ceria is also prepared by the combustion of cerous nitrate and ODH redox mixture. The theoretical equation for the formation of ceria is written as follows:

$$4Ce(NO_3)_3(aq) + 6C_2H_6N_4O_2(aq) + O_2(g)$$
$$\text{(ODH)}$$
$$\longrightarrow 4CeO_2(s) + 18N_2(g) + 12CO_2(g) + 18H_2O(g)$$
$$(12 \text{ mol of gases/mol of } CeO_2) \tag{4}$$

Though very fine CeO_2 is produced by cerous nitrate and ODH redox mixture, the reaction is highly exothermic and violent. This reaction should therefore be carried out in a bigger container with less than 3 g of $Ce(NO_3)_3$.

The actual weights of ceric ammonium nitrate and the different fuels used in the solution combustion method along with their particulate properties are listed in Table 5.2.

Table 5.2.   Compositions of redox mixtures and particulate properties of CeO_2.

Composition of redox mixtures	Powder density $(g\,cm^{-3})$	Surface area (m^2g^{-1})	Particle size[a] (nm)
$(NH_4)_2Ce(NO_3)_6$ (10 g) + 5.48 g GLY + 8.77 g of NH_4NO_3	4.1	14.0	105
$(NH_4)_2Ce(NO_3)_6$ (10 g) + (2.9 g) TFTA	4.0	52.0	29
$(NH_4)_2Ce(NO_3)_6$ (10 g) + (3.4 g) CH	3.9	65.0	27
$(NH_4)_2Ce(NO_3)_6$ (6.8 g) + (2.6 g) urea	3.9	84.0	19
$(NH_4)_2Ce(NO_3)_6$ (6.8 g) + (3.5 g) ODH	3.9	85.0	18

[a] From surface area.

Ceric ammonium nitrate–ODH redox mixtures have been employed for the preparation of ceria and noble metal-ion-substituted nano-ceria. In some cases glycine is also used.

The phase formation and fine particle nature of ceria is confirmed by powder XRD pattern which shows considerable line broadening (Fig. 5.3). The crystallite size of ceria calculated from X-ray line broadening varies from 20 to 35 nm. The particle size of ceria calculated from surface area is in the range of 18–30 nm.

5.3 SYNTHESIS OF METAL-ION-SUBSTITUTED CERIA

A higher catalytic activity can be expected if noble metal ions like Pd or Pt are substituted in the solid solution support such as $Ce_{1-x}Ti_xO_2$, because of additional electronic interaction between Ce^{4+}, Ti^{4+} and noble metal ions. Metal-substituted ceria ($Ce_{1-x}M_xO_{2-\delta}$ (M = Cu, Ag, Au, Pd, Pt, Rh), $Ce_{1-x}Zr_xO_2$ and $Ce_{1-x}Ti_xO_2$) are prepared by the combustion of aqueous solutions containing ceric ammonium nitrate, ODH, and corresponding metal salts like halides/nitrates.[5–12] The synthesis of Pd-or Pt-ion-substituted $Ce_{1-x}Ti_xO_2$ namely $Ce_{1-x-y}Ti_xPd_yO_{2-\delta}$ and $Ce_{1-x-y}Ti_xPt_yO_{2-\delta}$ is promising in this regard.[13,14] The exact weights of oxidizer and fuels used for the preparation of various metal-ion-substituted ceria are tabulated in Table 5.3.

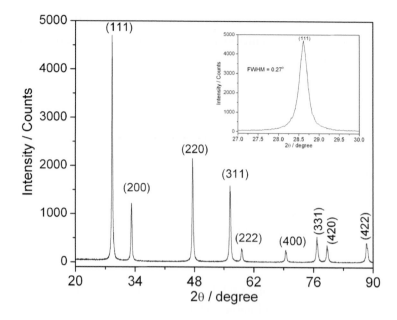

Fig. 5.3.   XRD pattern of CeO_2.

Table 5.3.   Compositions of redox mixtures for the preparation of $Ce_{1-x}M_xO_{2-\delta}$, $Ce_{1-x}Zr_xO_2$, $Ce_{1-x}Ti_xO_2$ and $Ce_{1-x}Ti(Pd/Pt)_xO_2$.

Composition of redox mixtures	Catalysts
10 g $(NH_4)_2Ce(NO_3)_6$, 0.293 g $Cu(NO_3)_2 \cdot 3H_2O$, and 4.227 g ODH	$Ce_{0.95}Cu_{0.05}O_{2-\delta}$
10 g $(NH_4)_2Ce(NO_3)_6$, 0.031 g $AgNO_3$, and 5.186 g ODH	$Ce_{0.99}Ag_{0.01}O_{2-\delta}$
10 g $(NH_4)_2Ce(NO_3)_6$, 0.063 g $HAuCl_4$, and 5.186 g ODH	$Ce_{0.99}Au_{0.01}O_{2-\delta}$
10 g $(NH_4)_2Ce(NO_3)_6$, 0.095 g H_2PtCl_6, and 5.175 g ODH	$Ce_{0.99}Pt_{0.01}O_{2-\delta}$
10 g $(NH_4)_2Ce(NO_3)_6$, 0.066 g $PdCl_2$, and 5.175 g ODH	$Ce_{0.98}Pd_{0.02}O_{2-\delta}$
10 g $(NH_4)_2Ce(NO_3)_6$, 4.75 ml of $RhCl_3 \cdot xH_2O$ (Arora Matthey Ltd. 40% Rh), and 5.175 g ODH	$Ce_{0.99}Rh_{0.01}O_{2-\delta}$
5 g $(NH_4)_2Ce(NO_3)_6$, 0.69 g $Zr(NO_3)_4$, and 2.1 g glycine	$Ce_{0.85}Zr_{0.15}O_2$
5 g $(NH_4)_2Ce(NO_3)_6$, 0.302 g $TiO(NO_3)_2$, and 1.955 g glycine	$Ce_{0.85}Ti_{0.15}O_2$
5 g $(NH_4)_2Ce(NO_3)_6$, 0.585 g $TiO(NO_3)_2$, 0.044 g $PdCl_2$, and 2.08 g glycine	$Ce_{0.73}Ti_{0.25}Pd_{0.02}O_{2-\delta}$
3 g $(NH_4)_2Ce(NO_3)_6$, 0.1823 g $TiO(NO_3)_2$, 0.0252 g $Pt(NH_3)_4(NO_3)_2$, and 1.1814 g glycine	$Ce_{0.84}Ti_{0.15}Pt_{0.01}O_{2-\delta}$

Assuming complete combustion, theoretical equations for the formation of metal-ion-substituted ceria ($Ce_{1-x}M_xO_{2-x/2}$) can be written as follows:

$$5(NH_4)_2Ce(NO_3)_6(aq) + 12C_2H_6N_4O_2(aq)$$
$$(ODH)$$

$$\xrightarrow{xCu(NO_3)_2/AgNO_3/HAuCl_4/H_2PtCl_6/PdCl_2/RhCl_3}$$

$$5Ce_{1-x}M_xO_{2-x/2}(s) + 44N_2(g) + 24CO_2(g) + 56H_2O(g)$$
$$(24.8 \text{ mol of gases/mol of } Ce_{1-x}M_xO_{2-x/2})$$

$$M^{n+} = Cu, Ag, Au, Pt, Pd, Rh; (n = 1 - 4) \qquad (5)$$

Similarly, chemical equations for the formation of $Ce_{1-x}(Zr/Ti)_xO_2$ can be written as follows:

$$9(1-x)(NH_4)_2Ce(NO_3)_6 + 9xZr(NO_3)_4$$
$$+ 20C_2H_5NO_2 + 4(1-x)C_2H_5NO_2$$
$$\longrightarrow 9Ce_{1-x}Zr_xO_2 + 40CO_2 + 8(1-x)CO_2$$
$$+ 28N_2 + 20(1-x)N_2 + 50H_2O + 46(1-x)H_2O \qquad (6)$$

$$9(1-x)(NH_4)_2Ce(NO_3)_6 + 9xTiO(NO_3)_2$$
$$+ 10C_2H_5NO_2 + 14(1-x)C_2H_5NO_2$$
$$\longrightarrow 9Ce_{1-x}Ti_xO_2 + 20CO_2 + 48(1-x)CO_2$$
$$+ 14N_2 + 34(1-x)N_2 + 25H_2O + 71(1-x)H_2O \qquad (7)$$

The various compositions of metal-ion-substituted ceria prepared by solution combustion route are summarized in Table 5.4.

Simultaneous substitution of two metals is achieved in a single step by the same method. The bimetal ionic catalysts are synthesized by the combustion of aqueous stoichiometric composition of 4.77 ml of 1% H_2PtCl_6 solution (1 g of H_2PtCl_6 in 100 ml of water), 2.37 ml of 1% $RhCl_3$ solution (1 g of $RhCl_3$ in 100 ml of water), 10 g ceric ammonium nitrate, and 5.18 g of oxalyl dihydrazide (ODH) as the fuel to give $Ce_{0.98}Pt_{0.01}Rh_{0.01}O_{2.8}$.[10]

Table 5.4.   Compositions of noble metal-ion-substituted ceria.

Metal ion	Oxide matrix		
	CeO_2	$Ce_{1-x}Ti_xO_2$	$Ce_{1-x}Zr_xO_2$
Cu^{2+}	$Ce_{1-x}Cu_xO_{2-\delta}$ ($x = 0.01$–0.1)	$Ce_{0.8-x}Ti_{0.2}Cu_xO_{2-\delta}$ ($x = 0.05, 0.1$)	—
Ag^+	$Ce_{1-x}Ag_xO_{2-\delta}$ ($x = 0.01$)	—	—
Au^{3+}	$Ce_{1-x}Au_xO_{2-\delta}$ ($x = 0.01$)	—	—
Pd^{2+}	$Ce_{1-x}Pd_xO_{2-\delta}$ ($x = 0.01, 0.02$)	$Ce_{0.75-x}Ti_{0.25}Pd_xO_{2-\delta}$ ($x = 0.01, 0.02$)	$Ce_{1-x-y}Zr_xPd_yO_{2-\delta}$ ($x = 0.15$–0.4; $y = 0.02$)
Rh^{3+}	$Ce_{1-x}Rh_xO_{2-\delta}$ ($x = 0.005, 0.01$)	—	—
Pt^{2+}	$Ce_{1-x}Pt_xO_{2-\delta}$ ($x = 0.01, 0.02$)	$Ce_{0.85-x}Ti_{0.15}Pt_xO_{2-\delta}$ ($x = 0.01, 0.02$)	—
$Pt^{2+} +$ Rh^{3+}	$Ce_{1-x}Rh_{x/2}Pt_{x/2}O_{2-\delta}$ ($x = 0.01, 0.02$)	—	—

5.4 CHARACTERIZATION OF METAL-ION-SUBSTITUTED CERIA

During combustion, metal salts added to the redox mixtures initially decompose to their respective metals and then get oxidized to their metal ions, forming ionically substituted ceria. The crystal structure, electronic, and catalytic properties of metal-ion-substituted ceria have been investigated by powder XRD patterns, Rietveld refinement, XPS, and TPR (temperature-programmed reduction). The ionic substitution is confirmed by powder XRD patterns and X-ray photoelectron spectroscopy, while the particle size is determined by X-ray line broadening and TEM studies.

All powder XRD patterns of metal-ion-substituted ceria are identical with the ceria pattern indicating the retention of fluorite structure. However, the ionic size of metal ions like Cu, Ag, Au, Pt, etc., being smaller, a decrease in the lattice parameter a is observed. The lattice parameter of pure CeO_2 is 0.5411 nm whereas 5% Cu-ion-substituted ceria has a value of 0.5409 nm. A decrease in the lattice parameter proves the ionic substitution. Typical powder XRD patterns of Pd-substituted ceria are shown in Fig. 5.4 and Rietveld refined XRD patterns are shown in Fig. 5.5.

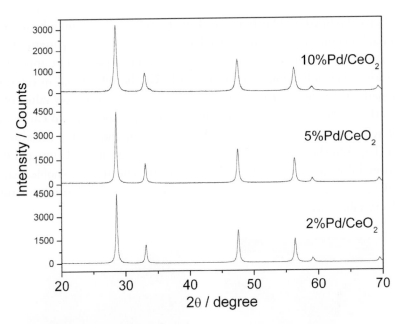

Fig. 5.4.   XRD patterns of as-prepared (a) 2% Pd/CeO₂ (b) 5% Pd/CeO₂, and (c) 10% Pd/CeO₂ catalysts.

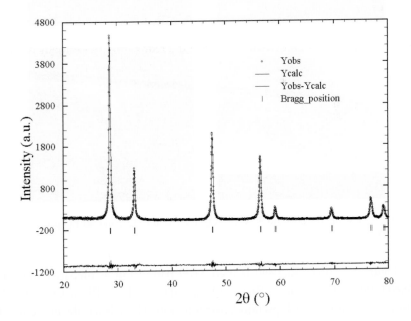

Fig. 5.5.   Rietveld refined XRD pattern of 2 atom% Pt/Ceria.

Rietveld refined XRD pattern clearly demonstrates Pd/Pt ion substitution in ceria. Lines due to either Pd/Pt metal or PdO are absent in the pattern. With increase in the metal content (more than 5%) metal peaks start appearing indicating that only about 3% Pd or 2% Pt can be substituted in ceria by this method. Single-phase $Ce_{1-x}Cu_xO_{2-\delta}$ is formed only for low values of x up to 0.07. There are no diffraction lines either corresponding to Cu or the weak reflections due to CuO or Cu_2O. TEM of ceria and Cu-ion-substituted ceria (Fig. 5.6) show that particle sizes are in the range of 25–35 nm and the electron diffraction is indexed to cubic fluorite. In general, 1–3% noble metal-ion (Cu, Ag, Au, Pt, Pd, Rh)-substituted ceria crystallizes in fluorite structure with crystallite sizes ranging from 20 to 25 nm.

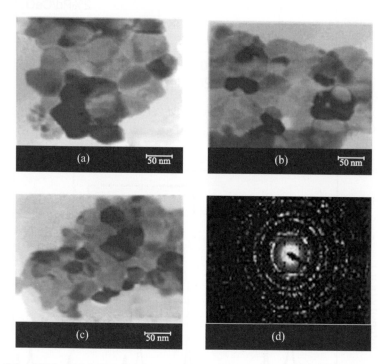

Fig. 5.6.   TEM of (a) CeO_2, (b) 3%Cu/CeO_2, (c) 5%Cu/CeO_2, and (d) ED pattern of 5% Cu/CeO_2.

Figure 5.7 shows Rietveld refined XRD patterns of as-prepared and calcined (800°C) $Ce_{0.99}Pt_{0.005}Rh_{0.005}O_{2-\delta}$ and $Ce_{0.99}Pt_{0.01}Rh_{0.01}O_{2-\delta}$ which display good fitting with pseudo-Voigt function. There is hardly any residual background indicating highly crystalline nature of the samples. Also no peaks are observed due to Pt or Rh metal or any of their oxides in the XRD pattern, clearly showing the formation of solid solution ($Ce_{0.99}Pt_{0.005}Rh_{0.005}O_{2-\delta}$). Pt and Rh ions are in $+2$ and $+3$ oxidation states, respectively, in this compound.

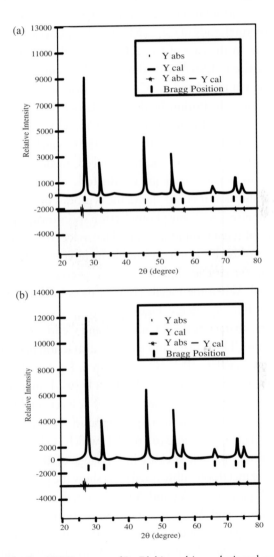

Fig. 5.7. Rietveld refined XRD pattern of Pt–Rh bimetal-ion-substituted ceria (a) as-prepared and (b) calcined (800°C).

Powder XRD pattern of $Ce_{1-x}Zr_xO_2$ and $Ce_{1-x}Ti_xO_2$ also shows retention of the fluorite structure demonstrating the formation of single phase with the X-ray diffraction lines being broader than pure ceria. The fluorite structure is retained with even up to 40% substitution of the Ti ion in the material. These crystallites do not have pores. Average crystallite sizes of ceria-zirconia and ceria–titania solid solutions are in the range of 15–25 nm.

Detailed structural study of $Ce_{1-x}M_xO_{2-\delta}$ has been carried out employing X-ray absorption. In the case of $Ce_{1-x}Cu_xO_{2-\delta}$, the Cu–O distance in the solid solution lattice is 1.96 Å with three coordination, indicating oxide ion vacancies around Cu ion. A unique Cu–O–Ce distance of 3.13 Å is observed confirming Cu ion substitution in the ceria lattice. It is important to note these details because stoichiometric oxide of Cu and Ce is not known in the literature. $Ce_{1-x}Cu_xO_{2-\delta}$ oxide with $x = 0.05, 0.1$ can be viewed as Cu-ion-substituted ceria with oxide ion vacancy. Indeed, oxide ion vacancy renders $Ce_{1-x}Cu_xO_{2-\delta}$ to be good oxide ion conductors.

Solution combustion method allows the synthesis of these oxide ion conductors in nanometric size. X-ray photoelectron spectroscopy predicts the electronic structure of a compound and provides exact oxidation states of ions. Oxidation states of both Ce ion and noble metal ions in the compound are determined from the core level photoelectron spectra of the component ions in the compound. When the metals are present as ions, their core levels are shifted to higher binding energy. Figure 5.8 shows the Cu (2p) and Ce (3d) core level spectra, characteristic of Cu in +2 and Ce in +4 states.[15]

High-resolution TEM image and electron diffraction patterns of combustion synthesized $Ce_{0.84}Ti_{0.15}Pt_{0.01}O_2$ is shown in Fig. 5.9. The fringes spaced at $\sim$3.20 Å corresponds to $Ce_{1-x}Ti_xO_2$ (111) layers. Fringes due to Pd or Pt metal particles separated at 2.30 Å are absent. The electron diffraction pattern clearly shows fluorite structure of the compound. EDAX analysis on these images reveal Pt:Ti:Ce ratio close to 76:23:1, which is close to the bulk composition. The average particle sizes are in the range of 20–15 nm.

XPS studies of metal-ion-substituted ceria show the presence of ionic metal atoms. The XPS patterns of Ag-and Au-substituted ceria show the presence of Ag in +1 state[16] and Au in +3 state. Some fraction of Au is dispersed as Au, i.e., atomic state.[7] Binding energy of Pd ($3d_{5/2}$) peak in Pd metal is observed at 335.0 eV; in PdO it is at 336.2 eV whereas the value is higher in $PdCl_2$ or $Pd(NO_3)_2$ at 338.2 eV, even though Pd is in +2 formal oxidation

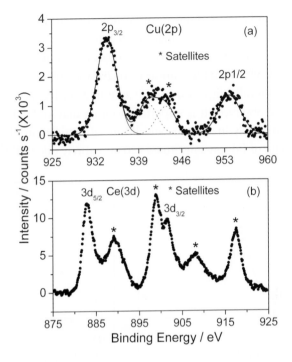

Fig. 5.8.   XPS of (a) Cu (2p) and (b) Ce (3d) levels in 5% Cu/CeO_2.

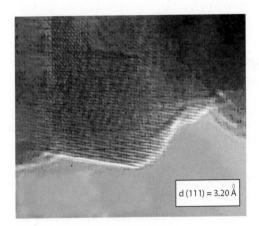

Fig. 5.9.   HRTEM image of $Ce_{0.84}Ti_{0.15}Pt_{0.01}O_2$.

state. Chloride and nitrate anions are more ionic than oxide ions, consequently charge state of Pd is higher than that in PdO. In Pd-ion-substituted ceria the binding energy is observed at 337.5 eV; as a result, Pd is in +2 state but more ionic than PdO. XPS of Pd (3d) core levels of 2% and 10% Pd in ceria are given in Fig. 5.10. Pd is fully dispersed as Pd^{2+} ions in 2% while in 10% sample Pd metal peaks are observed. It is important to note that in 2% Pd-substituted ceria Pd is completely in +2 state and Pd metal particles are absent. Hence, catalytic activity of Pd-ion-substituted ceria is entirely due to Pd^{2+} ion.[17]

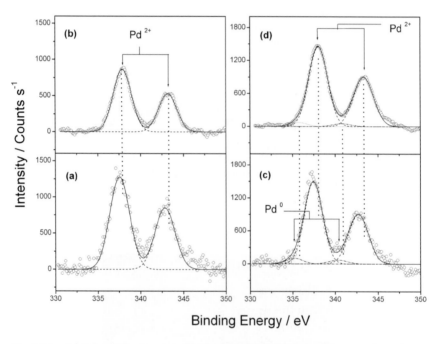

Fig. 5.10.   Pd (3d) core level spectra of (a) 2 atom% Pd/CeO_2 from ceric ammonium nitrate, (b) 2 atom% Pd/CeO_2 from cerous nitrate, (c) 5 atom% Pd/CeO_2 from ceric ammonium nitrate, and (d) 10 atom% Pd/CeO_2 from cerous nitrate.

X-ray photoelectron spectroscopy is the only direct method to show that Pt is in oxidized form in ceria (Fig. 5.11). Unlike Cu or Pd, multiple oxidation states are seen in Pt-substituted ceria. The spectra shown in the figure is complicated because Pt is present in +2 as well as +4 states. When they are resolved into +2 and +4 components it is found that over 80% of Pt is present in +2 state and about 12% Pt is present in +4 state.

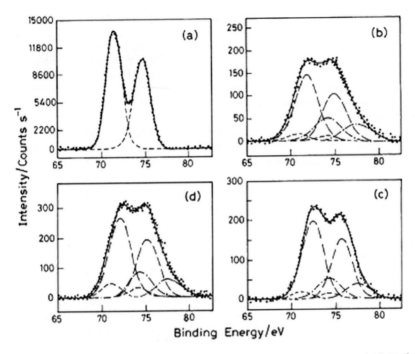

Fig. 5.11.   Pt (4f) XPS of (a) Pt metal, (b) 1% Pt/CeO$_2$, (c) 2% Pt/ CeO$_2$, and (d) 800°C heated 2% Pt/CeO$_2$.

XPS studies show that Cu, Pt, and Pd ions are doped in +2 state and Rh in Ce$_{1-x}$Rh$_x$O$_{2-\delta}$ is in +3 state. Therefore, for charge balancing there has to be oxide ion vacancy in metal-substituted ceria compounds. In addition, Ce itself can be partially present in +3 state to the extent of 5%.[18] Hence, the compound is written as Ce$_{1-x}$M$_x$O$_{2-\delta}$. The value of δ is slightly more than x.

Table 5.5 summarizes the binding energies of various Pt species present in Pt-, Rh-substituted ceria catalysts.

Dispersion of the noble metal component in ionic form over CeO$_2$ nanocrystallites can be an effective method to disperse Rh so that more of Rh ion can become active sites for adsorption, in contrast to Rh atoms on Rh metal particles supported on oxides. Stronger ionic interaction between Rh ion and CeO$_2$ can drastically reduce loss of noble metal by evaporation and sintering due to ageing of the catalyst.[9] Oxide ion vacancies present next to Rh ion have been confirmed from detailed EXAFS study. This is in contrast to Rh

Table 5.5.   Binding energies, FWHM's and relative intensities of different Pt species[a].

Catalyst	State	Pt species	Binding energy of $4f_{7/2}$ (eV)	FWHM (eV)	Relative intensity (%)
$Ce_{0.99}Pt_{0.005}$ $Rh_{0.005}O_{2-\delta}$	As-prepared	Pt^0	71.1	2.4	6.5
		Pt^{2+}	71.8	2.6	74
		Pt^{4+}	74.4	3.1	19.5
	Heated at 800°C	Pt^0	71.0	2.2	2
		Pt^{2+}	72.7	2.5	71
		Pt^{4+}	74.7	2.9	25
$Ce_{0.98}Pt_{0.01}$ $Rh_{0.01}O_{2-\delta}$	As-prepared	Pt^0	71.1	2.4	9.5
		Pt^{2+}	71.9	2.6	67
		Pt^{4+}	74.1	3.1	23.5
	Heated at 800°C	Pt^0	71.1	2.2	4.5
		Pt^{2+}	72.8	2.4	84.5
		Pt^{4+}	74.6	2.9	11

[a]Observed from the deconvoluted Pt (4f) spectra of Pt- and Rh-substituted ceria catalysts.

metal atoms in Rh metal particles where both oxidant and reductant molecules compete for the same metal atom.

Redox property of $Ce_{1-x}Rh_xO_{2-\delta}$ shows that Rh ion can be reduced from +3 to 0 valence state and therefore, one Rh ion can exchange three electrons. On the other hand, Cu^{2+}, Pt^{2+}, and Pd^{2+} ions in ceria can exchange only two electrons. Therefore, with lower concentration, Rh/ceria is a better redox catalyst.

5.5 OXYGEN STORAGE MATERIALS

The amount of oxygen that can be released under reducing condition and its uptake under oxidizing condition is called OSC. The origin of enhanced OSC in both $Ce_{1-x}Zr_xO_2$ and $Ce_{1-x}Ti_xO_2$ has recently been investigated in detail using powder XRD technique, EXFAS studies and TPR experiments.[11,12] Substitution of Zr or Ti in ceria enhances the OSC compared to pure ceria. TPR by H_2 is generally employed to estimate OSC. Even though ZrO_2 cannot be reduced by CO or H_2, OSC is enhanced in the $Ce_{1-x}Zr_xO_2$ solid solution.

This implies that Ce in solid solution can be reduced more easily compared to pure CeO_2. In $Ce_{1-x}Ti_xO_2$, both Ce^{4+} and Ti^{4+} ions can be reduced to $+3$ states and reducibility of the solid solution or OSC should be higher compared to both CeO_2 and TiO_2.

If noble metal ions like Pd or Pt are substituted in the solid solution support such as $Ce_{1-x}Ti_xO_2$, a higher catalytic activity ought to result because of additional electronic interaction between Ce^{4+}, Ti^{4+} and noble metal ions. The synthesis of Pd- or Pt-ion-substituted $Ce_{1-x}Ti_xO_2$ namely, $Ce_{1-x-y}Ti_xPd_yO_{2-\delta}$ and $Ce_{1-x-y}Ti_xPt_yO_{2-\delta}$, is attractive in this regard.[13,14] The results of this are discussed in the following sections.

OSC value is expressed in terms of volume or moles of hydrogen adsorbed per gram of ceria. For the nano-ceria prepared by solution combustion method the uptake of H_2 per gram is in the range of 10–30 cc. Actual values of OSC of CeO_2, $Ce_{1-x}Zr_xO_2$, $Ce_{1-x}Ti_xO_2$, $Ce_{1-x}Pd/Pt_xO_{2-\delta}$, and Pd-or Pt-substituted $Ce_{1-x}Ti_xO_{2-\delta}$ compounds reduced by hydrogen upto $675°C$ are summarized in Table 5.6.

Table 5.6. OSC of CeO_2, $Ce_{1-x}Zr_xO_2$, $Ce_{1-x}Ti_xO_2$, $Ce_{1-x}Pd/Pt_xO_{2-\delta}$ and Pd/Pt substituted $Ce_{1-x}Ti_xO_{2-\delta}$.

Compound	H_2 uptake up to $675°C$ $(mol/g) \times 10^4$	Formula of the reduced species	% of Ce^{4+} reduced to Ce^{3+}
CeO_2	2.3	$CeO_{1.96}$	8
TiO_2	9.9	$TiO_{1.92}$	—
$Ce_{0.9}Ti_{0.1}O_2$	7.4	$Ce_{0.9}Ti_{0.1}O_{1.88}$	15.5
$Ce_{0.85}Ti_{0.15}O_2$	8.8	$Ce_{0.85}Ti_{0.15}O_{1.86}$	15
$Ce_{0.8}Ti_{0.2}O_2$	11.2	$Ce_{0.8}Ti_{0.2}O_{1.83}$	17.5
$Ce_{0.8}Zr_{0.2}O_2$	29	$Ce_{0.8}Zr_{0.2}O_{1.79}$	—
$Ce_{0.7}Ti_{0.3}O_2$	13	$Ce_{0.7}Ti_{0.3}O_{1.77}$	22
$Ce_{0.7}Zr_{0.3}O_2$	17.2	$Ce_{0.7}Zr_{0.3}O_{1.73}$	—
$Ce_{0.6}Ti_{0.4}O_2$	20	$Ce_{0.6}Ti_{0.4}O_{1.73}$	23.5
$Ce_{0.6}Zr_{0.4}O_2$	23.6	$Ce_{0.7}Zr_{0.3}O_{1.64}$	—
$Ce_{0.5}Zr_{0.5}O_2$	25	$Ce_{0.5}Zr_{0.5}O_{1.63}$	—
$Ce_{0.99}Pt_{0.01}O_{2-\delta}$	2.9	$Ce_{0.99}Pt_{0.01}O_{1.95}$	10
$Ce_{0.98}Pd_{0.02}O_{2-\delta}$	1.8	$Ce_{0.99}Pd_{0.01}O_{1.97}$	6
$Ce_{0.84}Ti_{0.15}Pt_{0.01}O_{2-\delta}$	15.7	$Ce_{0.84}Ti_{0.15}Pt_{0.01}O_{1.75}$	36
$Ce_{0.73}Ti_{0.25}Pd_{0.02}O_{2-\delta}$	17	$Ce_{0.73}Ti_{0.25}Pd_{0.02}O_{1.75}$	21
$Ce_{0.83}Ti_{0.15}Pt_{0.02}O_{2-\delta}$	23.2	$Ce_{0.84}Ti_{0.15}Pt_{0.01}O_{1.63}$	60

Hydrogen uptake of ceria–titania solid solution is given in Fig. 5.12. It is observed that ceria–titania is as good as ceria–zirconia showing high OSC.

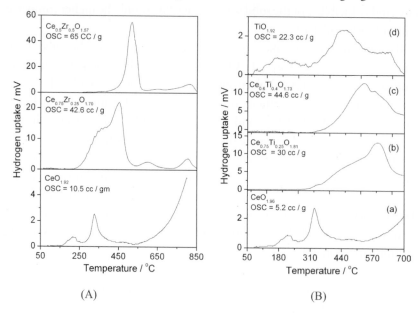

Fig. 5.12.   (A) H_2 uptake by (a) $CeO_{1.92}$, (b) $Ce_{0.75}Zr_{0.25}O_2$, and (c) $Ce_{0.5}Zr_{0.5}O_2$. (B) H_2 uptake by (a) $CeO_{1.96}$, (b) $Ce_{0.75}Ti_{0.25}O_{1.81}$, (c) $Ce_{0.6}Ti_{0.4}O_{1.73}$, and (d) $TiO_{1.92}$.

Due to the importance of OSC of ceria-titania and ceria–zirconia materials in auto exhaust catalysts, several studies have been carried out. The CO conversion versus temperature plot of $Ce_{0.84}Ti_{0.15}Pt_{0.01}O_{2-\delta}$ catalyst shows much lower light off temperature with $T_{50} = 150°C$ compared to $Ce_{0.99}Pt_{0.01}O_{2-\delta}$ with $T_{50} = 260°C$ (Fig. 5.13). Similarly, $Ce_{0.73}Ti_{0.25}Pd_{0.02}O_{2-\delta}$ shows higher catalytic activity with T_{50} at $110°C$ compared to $Ce_{0.98}Pd_{0.02}O_{2-\delta}$ with T_{50} at $135°C$.

NO reduction activity over $Ce_{0.84}Ti_{0.15}Pt_{0.01}O_{2-\delta}$ and $Ce_{0.99}Pt_{0.01}O_{2-\delta}$ is similar (Fig. 5.14). The rate of conversion at $135°C$ (at 15% conversion) is 2×10^{-7} mol g^{-1} s^{-1}. Activation energies are 23 and 24 kcal mol^{-1} for $Ce_{0.99}Pt_{0.01}O_{2-\delta}$, and $Ce_{0.84}Ti_{0.15}Pt_{0.01}O_{2-\delta}$ respectively. Thus, Ti-substituted ceria is a better oxide support than pure ceria for catalytic reactions. However, $Ce_{0.73}Ti_{0.25}Pd_{0.02}O_{2-\delta}$ shows high catalytic activity for NO reduction with 100% N_2 selectivity below $240°C$ which is lower than other catalysts reported in the literature.

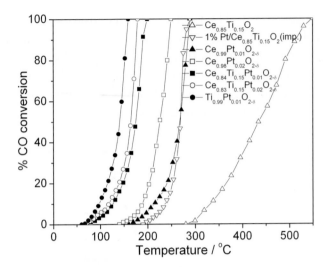

Fig. 5.13. %CO conversion as a function of temperature over $Ce_{0.85}Ti_{0.15}O_2$ and Pt, substituted oxides under the reaction condition: $CO = 2$ vol.%, $O_2 = 2$ vol.%, $F_t = 100$ sccm, GHSV= 43,000 h^{-1}, $W = 25$ mg.

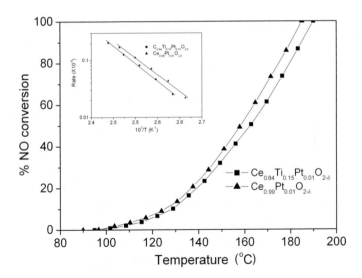

Fig. 5.14. % NO conversion versus temperature for the reaction NO + CO under the condition: NO = 0.25 vol.%, CO = 0.25 vol.%, $F_t = 40$ sccm, GHSV = 20,000 h^{-1}, $W = 150$ mg.

Light off curves of C_2H_2, C_2H_4, and C_3H_8 oxidation over 100 mg of the two catalysts show 1 vol.% "HC" is taken and "HC" to oxygen ratio is 1:5 with total flow of 100 sccm (Fig. 5.15). Clearly, $Ce_{0.84}Ti_{0.15}Pt_{0.01}O_{2-\delta}$ shows higher catalytic activity compared to $Ce_{0.99}Pt_{0.01}O_{2-\delta}$. The activation energies are lower for the $Ce_{0.84}Ti_{0.15}Pt_{0.01}O_{2-\delta}$ compared to $Ce_{0.99}Pt_{0.01}$ $O_{2-\delta}$ for oxidation of all the hydrocarbons studied here.

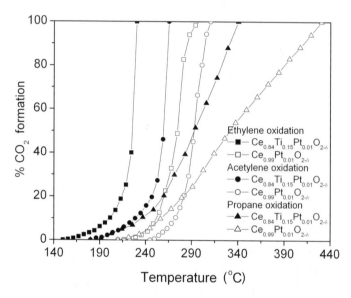

Fig. 5.15. (a) %CO_2 formation as a function of temperature for different hydrocarbon oxidation over $Ce_{0.99}Pt_{0.01}O_{2-\delta}$ and $Ce_{0.84}Ti_{0.15}Pt_{0.01}O_{2-\delta}$. Reaction condition: "HC" = 1 vol.%, O_2 = 5 vol.%, F_t = 100 sccm, GHSV= 43,000 h^{-1}, W = 100 mg.

It is inferred that the availability of free oxygen is increased by the transport of bulk oxygen to surface along with the movement of oxide ions from the tetrahedral sites to the vacant octahedral sites. However, a much higher OSC is obtained by the substitution of Zr/Ti ion for Ce in ceria forming $Ce_{1-x}Zr/Ti_xO_2$. Higher OSC of ceria–zirconia solid solution is shown to be a result of the destabilization of oxygen bonding with Ce and Zr ions in the lattice. Zr ion is bonded with four long Zr–O bonds (2.57 Å) and four short Zr–O bonds (2.15 Å) rather than the average Ce–O bond at 2.34 Å. The long bond being weaker, it is more easily removed by CO, hydrogen, or hydrocarbon giving higher OSC.

5.6 METAL-ION-SUBSTITUTED CERIA AS NANOCATALYSTS

The redox property of M ions in ceria matrix renders a catalytic character to $Ce_{1-x}M_xO_{2-\delta}$. The presence of noble metal ions in $Ce_{1-x}M_xO_{2-\delta}$ causes different redox properties than the corresponding MO or M_2O_3, due to the influence of the surrounding Ce ions in these compounds. This can be easily explained by a TPR test carried out with hydrogen on Cu-ion-substituted ceria.

From the TPR studies it is observed that Cu^{2+} ion in the solid solution, $Ce_{1-x}Cu_xO_{2-\delta}$, is reduced to Cu^0 at a temperature of $200°C$ when compared to Cu ion in CuO at $280–300°C$. In addition to the reduction of Cu ion to atomic Cu, two Ce^{4+} ions also get reduced to Ce^{3+} state as the oxygen connecting the metal ions in the crystal is removed. The reduced oxide can now be oxidized. Molecular oxygen gets dissociated and incorporates into the structure.[5] The redox reaction can be written as follows:

$$Ce_{1-x}Cu_xO_{2-\delta} + 2yH_2 \longrightarrow Ce_{1-x}Cu_xO_{2-\delta-2y} + 2yH_2O \qquad (8)$$

$$Ce_{1-x}Cu_xO_{2-\delta-2y} + yO_2 \longrightarrow Ce_{1-x}Cu_xO_{2-\delta} \qquad (9)$$

Such a reversible reduction–oxidation observed at $200°C$ means that yO_2 is the exchangeable oxygen, which makes Cu-ion-substituted ceria an oxidation catalyst. Thus, OSC property of ceria gets enhanced in terms of oxygen availability at a lower temperature than pure ceria. The fluorite structure of the catalyst remains intact during this redox reaction. The temperature at which oxygen exchange occurs depends on the metal ions. Lower the temperature of redox reaction, lower the catalytic oxidation temperature, and better is the oxidation catalyst.

As the metal ions are dispersed completely, larger numbers of sites for adsorption are created for pollutants like carbon monoxide. Similar to Cu, noble metal ions like Pt-, Pd-, Au-, and Ag- substituted ceria also act as oxidation catalysts. In fact they are better catalysts than Pt, Pd metal particles for the following reasons. Although the adsorption probability of carbon monoxide over Pt^{2+} ion is about the same as that on Pt atom on a Pt particle, in a nano-Pt metal particle only atoms on the surface of the particle are available for adsorption. Therefore, if Pt metal particles of 5 nm size are present then only about one-fifth of the total Pt atoms are available for adsorption. On the contrary, when Pt is dispersed as ions in an oxide matrix almost all the Pt atoms used for the preparation are available for adsorption as they are present mainly on the surfaces of nanocrystallites of $Ce_{1-x}Pt_xO_{2-\delta}$. This outcome results in five times more CO adsorption for the same amount of Pt. Additionally, electronic interaction between Pt ion and ceria causes enhancement of catalytic activity.

For the first time, it has been shown that Pt and Pd ions dispersed over ceria in the form of $Ce_{1-x}Pd/Pt_xO_{2-\delta}$ made by solution combustion display higher catalytic activity when compared to the same amount of Pt dispersed in metallic form.[8]

Reduction of metal oxide (MO) by CO is well known from the reactions:

$$MO + CO \rightarrow M + CO_2; \quad M = Cu, Pd, Pt \qquad (10)$$

$Cu + O_2$ however do not give CuO. Similarly, Pd and Pt metals are not oxidized by oxygen easily. Hence, CuO, PtO, and PdO do not act as oxidation catalysts. On the other hand, if Cu^{2+}, Pt^{2+}, or Pd^{2+} ions are substituted in the form of solid solution, $Ce_{1-x}M_xO_{2-x/2}$ (M = Cu, Pt, Pd) metal ions can be reduced to M^0 state. These get oxidized to +2 state to become catalysts. This is the thrust of the new generation catalyst developed by the solution combustion method. It is worthwhile to note that substitution of transition metals as well as ionic forms of noble metals in ceria could be achieved only by the solution combustion method. If the same synthesis is made by the co-precipitation method, prolonged heating is required that segregates metal ions; and merely by heating, noble metal oxides even if formed decompose to give metal particles. As a result, solution combustion method is unique to dope noble metal ions in ceria matrix.

Catalytic studies of all the metal-substituted ceria catalysts have been carried out using a completely computer-controlled TPR system equipped with a quadrupole mass spectrometer and an online gas chromatograph equipments (Fig. 5.16). Typical flow diagram of the set up is given in Fig. 5.17.

Metal-substituted ceria $Ce_{1-x}M_xO_{2-\delta}$ are found to be good catalysts for CO and HC (hydrocarbons) oxidation and NO reduction.

$$2NO + 2CO \longrightarrow 2CO_2 + N_2 \qquad (11)$$

$$2CO + O_2 \longrightarrow 2CO_2 \qquad (12)$$

$$\text{"HC"} + O_2 \longrightarrow H_2O + CO_2 \qquad (13)$$

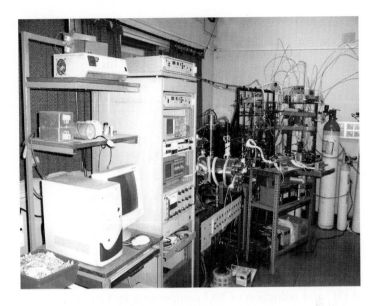

Fig. 5.16. Photograph of the TPR system.

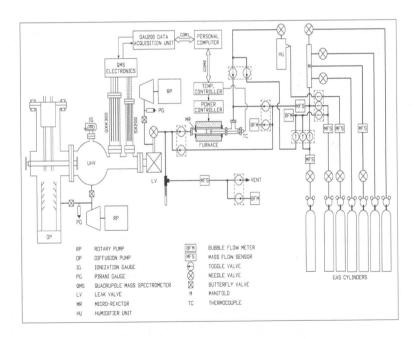

Fig. 5.17. Flow diagram of the TPR system.

CO, HC, and NO components are the major constituents found in vehicular exhaust. Their emission composition is given in Table 5.7.

Table 5.7.   Auto exhaust emission composition.

Emissions	Composition
CO	8500–10000 ppm
NO_x	800–1000 ppm
HC	500–600 ppm
Particulates	50–60
N_2	71.1 vol. %
CO_2	18 vol. %
H_2O	9.2 vol. %
O_2	7000–9000 ppm

A 100% conversion temperature observed for the oxidation reactions of CO and HC and reduction of NO components by $Ce_{1-x}M_xO_{2-\delta}$ catalysts in Eqs. (11)–(13) are summarized in Table 5.8.

Table 5.8.   100% conversion temperatures of NO, CO, and hydrocarbons reactions over $Ce_{1-x}M_xO_{2-\delta}$ catalysts.

Reactions	Cu/CeO_2 (°C)	Ag/CeO_2 (°C)	Au/CeO_2 (°C)	Pd/CeO_2 (°C)	Pt/CeO_2 (°C)
NO + CO	275	270	450	175	270
NO + NH_3	275	425	675	275	225
NO + CH_4	525	580	735	450	350
NO + C_3H_8	450	650	650	330	325
CO + O_2	275	260	350	175	180
CH_4 + O_2	450	550	600	330	400
C_3H_8 + O_2	350	400	425	230	110

$Ce_{0.95}Cu_{0.05}O_{2-\delta}$ is a good catalyst for CO oxidation and NO reduction. Cu^{2+} ion acts as an adsorption site for CO and the oxide ion vacancy created therein acts as a site for adsorption of oxygen and NO. The same compound is also an equally good catalyst for hydrocarbon oxidation. Overall, this compound is an improved TWC. Typical NO reduction profile is shown in Fig. 5.18. Redox potential of Cu is shown to be reduced by Ce ion in the $Ce_{1-x}Cu_xO_{2-\delta}$ due to electronic interaction between Cu and Ce ions.[5]

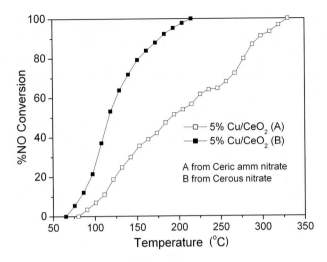

Fig. 5.18.   %NO conversion over Cu/CeO$_2$.

Ag and Au ions form a stable solid solution in ceria with 1–2 atom% substitution. Bulk gold is catalytically inactive due to its 5d^{10} electronic configurations and inability to donate electrons to oxygen, making oxidation by oxygen difficult. But, as the ionic radius of Au^{3+} is about 0.85 Å which is close to Ce^{4+} ion of 0.97 it is possible to substitute Au ion in ceria matrix.

Ag ion dispersed in ceria matrix as well as Ce$_{1-x}$Au$_x$O$_{2-\delta}$ shows catalytic activity for CO oxidation by oxygen and NO, as well as hydrocarbon oxidation by oxygen. Its catalytic activity is nearly as good as Cu-ion-substituted ceria. A comparison of catalytic activity for oxidation of CO over 1% Ag/Al$_2$O$_3$ or 1% Au/Al$_2$O$_3$ with 1 atom% Ag/CeO$_2$ or 1 atom% Au/CeO$_2$ shows that ionically dispersed Ag/Au in ceria is far superior to Ag/Au-metal-dispersed alumina.[6,7,19]

5.6.1 *Ce$_{1-x}$Pd$_x$O$_{2-\delta}$ as a Three-Way Catalyst*

Palladium-substituted ceria (Ce$_{0.98}$Pd$_{0.02}$O$_{2-\delta}$) is found to be extremely good for TWC reactions (Eqs. (11)–(13)) among the noble metal-substituted ceria including Cu^{2+} ions. The temperatures at which the CO oxidation, NO reduction, and hydrocarbon oxidation occur are the lowest for combustion synthesized catalysts when compared to similar catalysts reported in the literature. While many catalysts are known to reduce NO by CO most of the NO converts to N_2O. The Pd-substituted ceria gives over 80% N_2 selectivity and above 350°C shows 100% NO conversion to N_2. Figure 5.19 shows typical result of N_2 selectivity of this catalyst.[8] The reactivity of the compound is far higher than the corresponding amount of Pd metal loaded catalysts either on alumina or ceria.

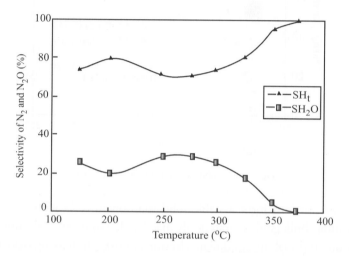

Fig. 5.19. N_2 and N_2O selectivity in NO + CO reaction over $Ce_{0.98}Pd_{0.02}O_{2-\delta}$.

This property of Pd-substituted ceria has been employed for coating cordierite honeycomb monolith in a single step to study its performance as a TWC.[20] For automobile applications, the ceramic monolith support with a wall structure like a honey comb is first wash coated with high surface area oxide like γ-Al_2O_3. On this substrate, the TWC coat is applied by sol–gel or slurry coating methods.[21] Presently, coating of Pd-substituted ceria on cordierite monolith is achieved in a single step by solution combustion of ceric ammonium nitrate, ODH, and $PdCl_2$ redox mixture at 500°C.[20] A uniform coating is achieved at much lower temperature than normally used and its adhesion on the cordierite surface is very strong (Fig. 5.20).

The weight of active catalyst in the matrix varies from 0.02% to 2 wt.% which is sufficient for catalytic activity but it can be loaded even up to 12 wt.% by repeated dip dry combustion. About 100% CO conversion is achieved below 80°C at a space velocity (SV) of 880 h^{-1}. At the same space velocity, 100% NO conversion is achieved below 185°C and 100% conversion of "HC"(C_2H_2) is achieved below 220°C. The three-way catalytic activity of

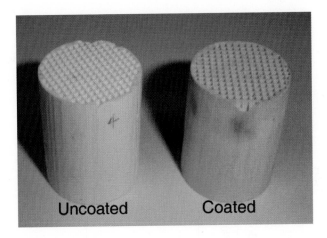

Fig. 5.20. Uncoated and Pd-substituted ceria coated cordierite monolith.

this catalyst shows that 100% conversion of all the pollutants occurs below 220°C with 15% excess oxygen (Fig. 5.21).

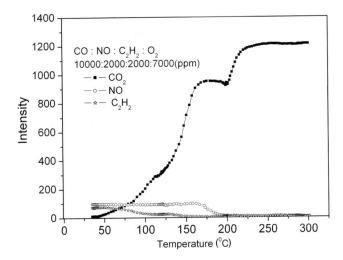

Fig. 5.21. Three-way catalytic performance over monolith for 10,000 ppm of CO, 2000 ppm of NO, 2000 ppm of C_2H_2 in the presence of 7000 ppm of O_2.

The high catalytic activity for Pd-substituted ceria for CO oxidation reaction is reflected by the comparative data given in Table 5.9. The small amount

Table 5.9.   Comparison of catalytic activity of metal-substituted ceria with other monolith catalyst.

Catalyst	Metal loading (g L^{-1})	Washcoat loading (g L^{-1})	[CO] (Vol.%)	[O] (Vol.%)	SV (h^{-1})	T_{50} (°C)	T_{100} (°C)
Pd/Al$_2$O$_3$	1.1	110	0.51	0.73	50,000	275	310 (100%)
Pd-Rh/Al$_2$O$_3$	1.32	110	0.51	0.73	50,000	240	250 (90%)
Pd/Al$_2$O$_3$–CeO$_2$, Pd-Rh/Al$_2$O$_3$	1.1	110	0.51	0.73	50,000	200	270 (100%)
CeO$_2$	1.32	110	0.51	0.73	50,000	220	260 (100%)
Pd/Al$_2$O$_3$	0.17	81.4	0.1	10	39,000	188	195 (100%)
Ce$_{0.98}$Pd$_{0.02}$O$_{2-\delta}$	0.23	18.1	0.16	0.16	21,000	170	220 (100%)

of Pd catalyst required for the reaction clearly demonstrates its efficiency over other monolith catalysts.

The higher catalytic activity results because of the ionic interaction between Pd and Ce ions via oxide ion and oxide ion vacancy. The ionic interaction leads to strong metal ion–ceria contact. The mechanism given in Fig. 5.22 takes into account the adsorption of CO on Pd ion, while oxygen is adsorbed on the oxide ion vacant sites. The oxygen molecule dissociates on the oxide ion vacant site by receiving electrons from CO through Pd ion. Since the CO is at the atomic distance with oxygen, formation of CO_2 is facilitated. A similar mechanism is proposed for the reduction of NO by CO. High selectivity of N_2 is attributed to the dissociative chemisorptions of NO over oxide ion vacancy.

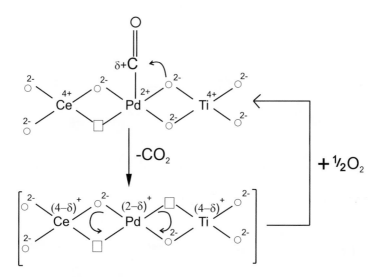

Fig. 5.22.   Schematic diagram of CO oxidation mechanism by oxygen.

Several advantages of single step coating of cordierite honeycomb monolith are

- Low cost loading of the precious metal, i.e., less amount of precious metal used.
- Short time for overall preparation.
- Low temperature required for the conversion of all the three pollutants.
- Avoiding risk of handling nanopowders.

5.6.2 $Ce_{1-x}Pt_xO_{2-\delta}$

It is well known that on Pt metal, CO is adsorbed molecularly and oxygen dissociatively leading to CO oxidation by Langmuir–Hinshelwood mechanism (Fig. 5.23). When Pt metal ions in +2 state are substituted in ceria forming a solid solution $Ce_{1-x}Pt_xO_{2-\delta}$, it forms an active site for CO adsorption and creates an oxide ion vacancy which acts as an adsorption center for the oxygen molecule. Since Pt is fixed in ionic form, it does not evaporate even at temperatures up to 800°C. In this way, higher catalytic activity is realized due to 100% Pt dispersion. $Ce_{1-x}Pt_xO_{2-\delta}$ catalysts are extremely good for CO and hydrocarbon oxidation and NO reduction.[9,22,23] It is a good oxidation catalyst and is stable in acid as well as alkali solutions. Therefore, the catalyst shows

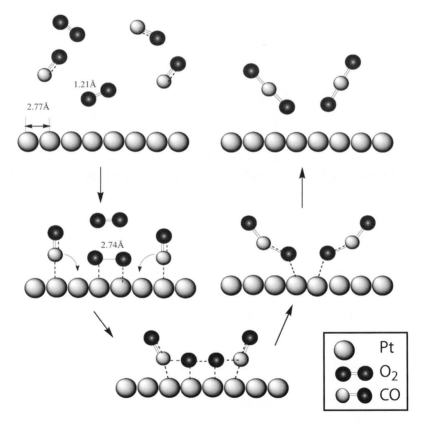

Fig. 5.23.   The Langmuir–Hinshelwood mechanism.

high electrocatalytic activity. For most of the applications 2 atom% Pd/Pt in ceria is more than sufficient for catalysis. Pt-ion-substituted ceria works better than Pt metal thereby opening up new vistas in electrocatalysis.

One of the extraordinary properties of $Ce_{1-x}Pt_xO_{2-\delta}$ is its hydrogen adsorption. Generally, hydrogen is known to be dissociatively chemisorbed on Pt metal atoms at room temperature; and one hydrogen atom is adsorbed on every Pt atom on the surface of the Pt metal particle. However, when Pt metal particles are dispersed on alumina, more than one hydrogen atom per Pt atom is observed and this phenomenon in catalysis is called "hydrogen spill-over." Even at a temperature of 200°C, H/Pt atoms ratio is about 1.5–2 per Pt atom. With $Ce_{1-x}Pt_xO_{2-\delta}$ oxide, the number of hydrogen atoms per Pt ion is found to vary with Pt ion concentration (x) in the compound as seen in temperature-programmed hydrogen adsorption (Fig. 5.24). The molar ratio of H_2 to Pt as a function of moles of Pt in ceria is found to be 2.5 (H/Pt = 5) for 1 atom% Pt in ceria at 0°C (Fig. 5.25). The H_2/Pt ratio decreases with increase in Pt content. Contrary to this, the H_2/Pt ratio for adsorption of hydrogen over Pt nanoparticles is only 0.078. This indicates that over 32 times more hydrogen is adsorbed on the ionically substituted Pt in ceria. The property of creating oxide ion vacancies in ceria by Pt in +2 state is crucial in achieving this result. It was found by serendipity that this catalyst combines hydrogen and oxygen at room temperature forming water without explosion.[24] This property has been exploited for application in valve regulated lead acid (VRLA) batteries with long life of 15–20 years.[25]

Combustion synthesized $Ce_{1-x}Pt_xO_{2-\delta}$ catalyst is also a superior catalyst for water gas shift reaction (WGS) to manufacture hydrogen:

$$CO + H_2O \longrightarrow H_2 + CO_2 \qquad (14)$$

CO conversion is maximum at 200°C over $Ce_{1-x}Pt_xO_{2-\delta}$ catalysts without any methanation reaction. The reaction rates are 1.86 and 4.66 $\mu mol\,g^{-1}\,s^{-1}$ at 125°C and 150°C, respectively, with a dry gas flow rate of $6\,L\,h^{-1}$ over 2% Pt/CeO_2.[23]

5.6.3 $Ce_{1-x}Rh_xO_{2-\delta}$

Rh/CeO_2 also shows unique hydrogen reduction–oxygen oxidation properties. H_2/Rh ratio is more than 5 with 0.5% Rh in ceria, and it is 2.5 with 1% Rh in

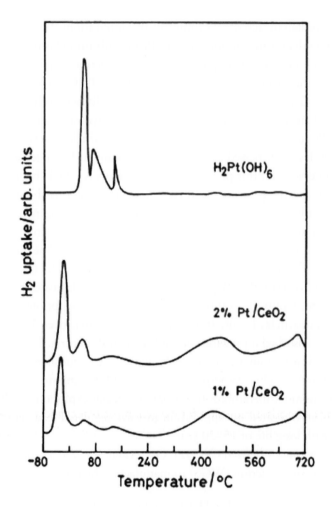

Fig. 5.24.   TPR of H_2 over $Ce_{1-x}Pt_xO_{2-\delta}$.

ceria. As each of the Rh ion in $+3$ state is reduced to zero valence state the Ce ion also gets reduced to $+3$ state. As a result, the H_2/Rh ratio is more than 1.5 in $Rh–CeO_2$. The TPR plots show Rh ion reducing at a lower temperature than Rh_2O_3 (Fig. 5.26).

$Ce_{1-x}Rh_xO_{2-\delta}$ is a good CO oxidation and NO reduction catalyst. Rates of CO oxidation, NO reduction, and hydrocarbon oxidation are 10–20 times higher than Rh metal dispersed over ceria and alumina for same amount of Rh. The oxide ion vacancy created due to Rh ion substitution acts as an oxygen

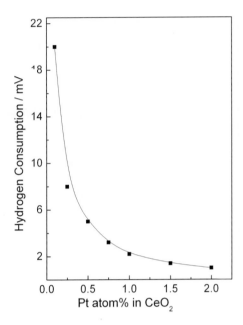

Fig. 5.25.   Plot of hydrogen consumption versus Pt moles.

adsorption center. Thus, in $Ce_{1-x}Rh_xO_{2-\delta}$, Rh ions are adsorption sites for molecules such as CO, hydrogen, and ammonia, hydrocarbon while the oxide ion vacancies are adsorption sites for oxygen, nitric oxide, and so on. A 100% dispersion and electronic interaction between Rh ion and Ce ion via oxide ion and oxide ion vacancy are some of the reasons for higher catalytic activity of $Ce_{1-x}Rh_xO_{2-\delta}$ vis-a-vis Rh metal over alumina.

Thus, almost all the transition metal ions substituted for Ce ion gets reduced at a lower temperature compared to the corresponding metal oxides, indicating ceria–metal ion interaction. The redox properties of metal ions are thus modified due to their presence in the ceria matrix. The reversible reduction/oxidation properties at a lower temperature make the noble metal-ion-doped ceria a better catalyst than the nano-metal particles dispersed over alumina, silica.

5.6.4 *Bimetal Ionic Catalysts* ($Ce_{1-x}Pt_{x/2}Rh_{x/2}O_{2-\delta}$)

Bimetal catalysts in the form of Pt–Pd, Pt–Ru, Pt–Rh, and Ni–Pd alloys are shown to give higher catalytic properties compared to monometal catalysts.

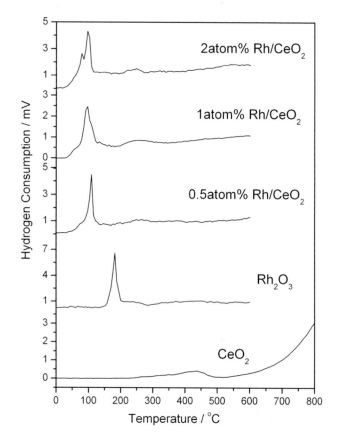

Fig. 5.26.   TPR of (a) CeO_2, (b) Rh_2O_3, and (c)–(d) 0.5–2% Rh/CeO_2.

For example, Pt- and Rh-ion-substituted ceria showed higher catalytic activity than either Pt- or Rh-substituted monometal ionic catalysts.

Catalytic activity of bimetallic nanoparticles in ceria is attributed to synergistic effect, where two metals such as 0.5% Pt and 0.5% Rh in ceria is found to be better than 1% Pt in ceria or 1% Rh in ceria.[10] Activity of bimetal ionic catalyst when compared with the corresponding monometal ionic catalysts shows the following features:

(i) Low-temperature reduction of Rh^{3+} ion in the bimetal ionic catalysts compared to monometal ionic $Ce_{1-x}Rh_xO_{2-\delta}$ catalyst.

(ii) Synergistic effect of bimetal Pt^{2+}, Rh^{3+} ion catalyst is shown to be due to Rh^{3+}–Rh^0 redox behavior at lower temperature in the presence of Pt^{2+} ion.

(iii) Rates of CO and C_2H_4 oxidation by O_2 and NO reduction by CO over bimetal ionic $Ce_{1-x}Pt_{x/2}Rh_{x/2}O_{2-\delta}$ are higher than corresponding monometal ionic $Ce_{1-x}Pt_xO_{2-\delta}$ and $Ce_{1-x}Rh_xO_{2-\delta}$ catalysts.

5.7 CONCLUDING REMARKS

Nano-ceria and metal-ions-substituted nano-ceria $Ce_{1-x}M_xO_{2-\delta}$ (M = Pt, Pd, Rh) prepared by solution combustion method exhibit better redox property and OSC leading to unique catalytic properties. The active metal content in these nanocatalysts being in small atom% range, there is considerable cost saving in making them. These properties have been exploited for application in

- VRLA batteries with long life.
- Hydrogen manufacture from low-temperature water gas shift reaction.
- TWC for automobile exhaust.

References

1. Taylor KC, Automobile catalytic converters, in Anderson JR, Boudart M (eds.), *Catalysis Science and Technology*, Vol. 5, Springer Verlag, Berlin, 1984.
2. Trovarelli A, *Catalysis by Ceria and Related Materials*, Imperial College Press, UK, 2002.
3. Maria Amala Sekar M, Sundar Manohar S, Patil KC, Combustion synthesis of fine particle ceria, *J Mater Sci Lett* **9**: 1205–1206, 1990.
4. Mimani T, Patil KC, Solution combustion synthesis of nanoscale oxides and their composites, *Mater Phys Mech* **4**: 134–137, 2001.
5. Bera P, Aruna ST, Patil KC, Hegde MS, Studies on Cu/CeO$_2$: A new NO reduction catalyst, *J Catal* **186**: 36–44, 1999.
6. Bera P, Patil KC, Hegde MS, NO reduction and CO and hydrocarbon oxidation over Ag/CeO$_2$, *Phys Chem Chem Phys* **2**: 3715–3719, 2000.
7. Bera P, Hegde MS, Characterizarion and catalytic properties of combustion synthesized Au/CeO$_2$ catalysts, *Catal Lett* **79**: 75–81, 2002.
8. Bera P, Patil KC, Jayaram V, Subbanna GN, Hegde MS, Ionic dispersion of Pt and Pd on CeO$_2$ by combustion method: Effect of metal ceria interaction

on catalytic activities for NO reduction and CO oxidation and hydrocarbon oxidation, *J Catal* **196**: 293–301, 2000.

9. Gayen A, Priolkar KR, Sarode PR, Jayaram V, Hegde MS, Emura S, $Ce_{1-x}Rh_x$ $O_{2-\delta}$ solid solution formation in combustion synthesized Rh/CeO_2 catalyst studied by XRD, TEM, XPS and EXAFS, *Chem Mater* **16**: 2317–2328, 2004.

10. Gayen A, Baidya T, Biswas K, Roy S, Hegde MS, Synthesis, structure and three way catalytic activity of $Ce_{1-x}Pt_{x/2}Rh_{x/2}O_{2-\delta}$ ($x = 0.01$ and 0.02) nano crystallites: Synergistic effect in bimetal ionic catalysts, *Appl Catal A Gen* **315**: 135–146, 2006.

11. Dutta G, Vagmare U, Baidya T, Hedge MS, Priolkar KR, Sarode PR, Reducibility of $Ce_{1-x}Zr_xO_2$: Origin of enhanced oxygen storage capacity, *Catal Lett* **108**: 165–172, 2006.

12. Dutta G, Vaghmare U, Baidya T, Hedge MS, Priolkar KR, Sarode PR, Origin of enhanced reducibility/oxygen storage capacity of $Ce_{1-x}Ti_xO_2$ compared CeO_2 or TiO_2, *Chem Mater* **18**: 3249–3256, 2006.

13. Baidya T, Marimuthu A, Hegde MS, Ravishankar N, Giridhar Madras, Higher catalytic activity of nano-$Ce_{1-x-y}Ti_xPd_yO_{2-\delta}$ compared to nano-$Ce_{1-x-y}Ti_xPt_yO_{2-\delta}$ for CO oxidation and N_2O and NO reduction by CO: Role of oxide ion vacancy, *J Phys Chem* C **111**: 830–839, 2007.

14. Baidya T, Gayen A, Hegde MS, Ravishankar, Loic Dupont, Enhanced reducibility of $Ce_{1-x}Ti_xO_2$ compared to that of CeO_2 and higher redox catalytic activity of $Ce_{1-x-y}Ti_xPt_yO_{2-\delta}$ compared to that of $Ce_{1-x}Pt_xO_2$, *J Phys Chem* B **110**: 5262–5272, 2006.

15. Bera P, Priolkar KR, Sarode PR, Hegde MS, Emura S, Kumashiro R, Lalla NP, Structural investigation of combustion synthesized Cu/CeO_2 catalysts by EXAFS and other physical techniques: Formation of a $Ce_{1-x}Cu_xO_{2-\delta}$ solid solution, *Chem Mater* **14**: 3591–3601, 2002.

16. Sarode PR, Priolkar KR, Bera P, Hegde MS, Emura S, Kumashiro R, Study of local environment of Ag in Ag/CeO_2 catalyst by EXAFS, *Mater Res Bull* **37**: 1679–1690, 2002.

17. Priolkar KR, Bera P, Sarode PR, Hegde MS, Emura S, Kumashiro R, Lalla NP, Formation of $Ce_{1-x}Pd_xO_{2-\delta}$ solid solution in combustion synthesized Pd/CeO_2 catalyst: XRD, XPS and EXAFS investigation, *Chem Mater* **14**: 2120–2128, 2002.

18. Bera P, Priolkar KR, Arup Gayen, Sarode PR, Hegde MS, Emura S, Kumashiro R, Jayaram V, Subbanna GN, Ionic dispersion of Pt over CeO_2 by combustion method: Structural investigation by XRD, TEM, XPS and EXAFS, *Chem Mater* **15**: 2049–2060, 2003.

19. Bera P, Patil KC, Jayaram V, Hegde MS, Subbanna GN, Combustion synthesis of nanometals supported on α-Al$_2$O$_3$; CO oxidation and NO reduction catalysis, *J Mater Chem* **9**: 1801–1805, 1999.

20. Sharma S, Hegde MS, Single step direct coating of 3-way catalysts on cordierite monolith by solution combustion method: High catalytic activity of Ce$_{0.98}$Pd$_{0.02}$O$_{2-\delta}$, *Catal Lett* **112**: 69–75, 2006.

21. Nijhuis TA, Beers AEW, Vergunst T, Hock I, Kapteijn F, Moulijn JA, Preparation of monolithic catalysts, *Catal Rev* **43**: 345–380, 2001.

22. Bera P, Arup Gayen, Hegde MS, Lalla NP, Spadaro L, Frusteri F, Arena F, Promoting effect of CeO$_2$ in combustion synthesized Pt/CeO$_2$ catalyst for CO oxidation, *J Phys Chem* B **107**: 6122–6130, 2003.

23. Bera P, Malwadkar S, Gayen A, Satyanarayana CVV, Rao BS, Hegde MS, Low temperature water gas shift reaction on combustion synthesized Ce$_{1-x}$Pt$_x$O$_{2-\delta}$ catalyst, *Catal Lett* **96**: 213–219, 2004.

24. Bera P, Hegde MS, Patil KC, Combustion synthesized Ce$_{1-x}$Pt$_x$O$_{2-\delta}$ ($x = 0.005, 0.01$ and 0.02, $\delta \sim 0.07$ and 0.1: A novel room temperature H$_2$–O$_2$ recombination catalyst, *Curr Sci* **80**: 1576–1578, 2001.

25. Hariprakash B, Bera P, Martha SK, Gaffoor SA, Hegde MS, Shukla AK, Ceria-supported platinum as hydrogen–oxygen recombinant catalyst for sealed lead–acid batteries, *Electrochem Solid-State Lett* **4**: A23–A26, 2001.

Chapter 6

Nanocrystalline Fe_2O_3 and Ferrites

6.1 MAGNETIC MATERIALS

Ferromagnetic materials that are mainly composed of ferric oxide (Fe_2O_3) are called "Ferrites." Magnetite (Fe_3O_4), a natural mineral is a genuine ferrite. Based on the crystal structure, ferrites have been categorized as spinels, perovskites, garnets, and hexaferrites (magnetoplumbite). They are also classified as soft, hard, microwave, and square loop depending on their magnetic performance (M–H curve). The technical applications of ferrites depend on the combination of intrinsic properties such as saturation magnetization, Curie temperature, microstructure, porosity, grain size, etc. Commercially important ferrites should possess high purity, chemical homogeneity, fine grain size, and high density.

Nanocrystalline magnetic oxides display unique phenomena such as superparamagnetism, magneto-optic, magneto-caloric, GMR, etc. These exceptional micromagnetic properties make them technologically important. Some of the applications of nanocrystalline magnetic oxides are in the miniaturization of devices such as motors, transformers, sensors, inductors, insulators, and information storage. Other applications include ferro fluids, perpendicular recording, stealth paint (bombers), low-temperature sintering to achieve high-density materials and catalysts.[1] Magnetic nanoparticles are also being increasingly used in biomedical engineering today. They are used as contrast media in radiology for the detection of tumors by magnetic resonance imaging (MRI). Magnetic hyperthermia is being explored as a possible alternative treatment of cancer. Ferrite systems like $Co_{1-x}Zn_xFe_2O_4$ and $Y_3Fe_{5-x}Al_xO_{12}$ with tuned T_c are the materials for this purpose.

The utilization of magnetic polymers to rapidly isolate labile enzymes or enzyme complexes after their liberation from the cell is another use.[2] Several applications of different ferrites are summarized in Table 6.1.

Table 6.1.   Ferrites and their applications.

Ferrites	Structure	Applications
γ-Fe_2O_3	Defect spinel	Audio/video magnetic tapes, cancer detection and treatment
Fe_3O_4	Spinel	Toner in photocopier
MFe_2O_4	Spinel	Transformer cores or inductors, TV deflector yokes
$Co_{1-x}Zn_xFe_2O_4$	Spinel	Cancer treatment
Mixed ferrite (Li–Zn, Mg–Zn, Ni–Zn, Mn–Zn)		Transformer and inductor cores
$LnFeO_3$	Perovskite	Read write heads, microwave and bubble domain devices
$Ln_3Fe_5O_{12}$	Garnet	Microwave applications, TV deflector tubes
$Ba/SrFe_{12}O_{19}$ (hexaferrites)	Magnetoplumbite	Permanent magnets, relays, loud speakers, recording tapes and discs, and floppy discs

So far, the clinical use of magnetic nanoparticles has focused on iron oxide. This is because of the chemical stability, biological compatibility, and relative ease of manufacture of magnetite (Fe_3O_4) and maghemite (γ-Fe_2O_3) nanoparticles. When produced in nanoparticulate form, both Fe_3O_4 and Fe_2O_3 exhibit superparamagnetic behavior at room temperature. In other words, they magnetize strongly under an applied magnetic field but retain no permanent magnetism once the field is removed. Both magnetite and maghemite have been approved as MRI contrast agents for clinical use. MRI agents work by altering the relaxation rates of water protons that try to realign with the static magnetic field following application of radiofrequency (RF) pulses. Iron oxide based contrast agents affect only transverse relaxation times (T2 decay) and not longitudinal relaxation (T1 decay). This leads to negative contrast or dark spots on T2-weighted MR images.[2]

Nanoparticle ferrites are usually obtained by wet chemical methods such as spray drying, freeze-drying, and co-precipitation.[1] Nanocrystalline spinel ferrites synthesized by combustible solid precursor method have been discussed in Chap. 2. Presently, the preparation of nano-Fe_2O_3 and ferrites by solution combustion method is described.[3–5] Their properties are compared with those obtained by the combustible solid precursors.

6.2 γ-Fe$_2$O$_3$

γ-Fe$_2$O$_3$, is the most widely used recording material in magnetic tapes and disks. The requirement for such materials is that they must have high coercivity, saturation magnetization, and remanance ratio. Substitution of cobalt ions for iron leads to a great increase in the coercivity of the final magnetic media.

γ-Fe$_2$O$_3$ is prepared by the combustion of aqueous solutions containing stoichiometric amounts of ferric nitrate and ODH/MDH.[6] In a typical experiment an aqueous solution containing 10.0 g Fe(NO$_3$)$_3$ · 9H$_2$O and 3.06 g of MDH (C$_3$H$_8$N$_4$O$_2$) in the molar ratio of 1.0:0.94, is rapidly heated at $350 \pm 10°$C. It ignites to yield fine-particle γ-Fe$_2$O$_3$ in a single step. The theoretical equations for the formation of γ-Fe$_2$O$_3$ are given below.

$$2Fe(NO_3)_3(aq) + \underset{\text{ODH}}{3C_2H_6N_4O_2(aq)}$$

$$\xrightarrow{350°C/1\,atm} Fe_2O_3\,(s) + 6CO_2(g) + 9N_2(g) + 9H_2O(g)$$

$$(24\,\text{mol of gases/mol of Fe}_2O_3) \qquad (1)$$

$$16Fe(NO_3)_3(aq) + \underset{\text{MDH}}{15C_3H_8N_4O_2(aq)}$$

$$\xrightarrow{350°C/1\,atm} 8Fe_2O_3(s) + 45CO_2(g) + 54N_2(g) + 60H_2O(g)$$

$$(\sim20\,\text{mol of gases/mol of Fe}_2O_3) \quad (2)$$

Formation of γ-Fe$_2$O$_3$ is confirmed by powder XRD patterns. It is observed that XRD pattern of the powder obtained by ODH process shows the presence of trace amounts of α-Fe$_2$O$_3$ along with γ-Fe$_2$O$_3$ whereas the powder obtained from MDH shows exclusively the presence of γ-Fe$_2$O$_3$ (Fig. 6.1). This appears to be due to the high exothermicity of the ODH process (Eq. (1)) which transforms some of the formed γ-Fe$_2$O$_3$ to α phase. When MDH is used as a fuel (Eq. (2)) large amount of gases evolve, leading to cooling and a lower exothermicity. The XRD pattern shows considerable line-broadening, indicating the fine-particle nature of the γ-Fe$_2$O$_3$. Crystallite size calculated from X-ray line broadening is in the range of 15–20 nm.

The variations of the magnetization moment and coercivity as a function of firing temperature are shown in Fig. 6.2. It can be seen that as the firing temperature is increased the saturation moment decreases and coercivity increases.

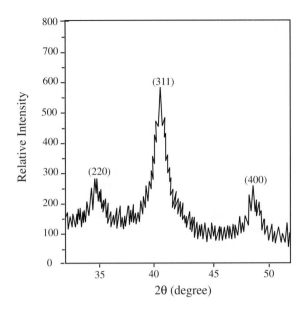

Fig. 6.1.   XRD pattern of γ-Fe₂O₃.

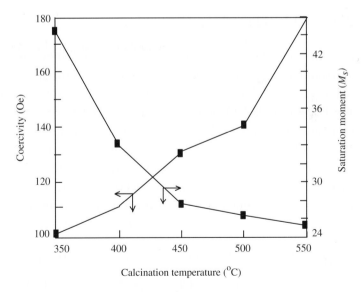

Fig. 6.2.   Plot of saturation moment and coercivity as a function of the calcination temperature of γ-Fe₂O₃.

The particulate and magnetic properties of γ-Fe$_2$O$_3$ prepared by the MDH process are comparable with those obtained by the combustible precursors (Table 6.2).

Table 6.2.   Comparison of properties of γ-Fe$_2$O$_3$ prepared by solution combustion and solid combustible precursor methods.

Property	Combustion method	Precursor method
Lattice constant (nm)	0.837	0.837
Crystallite size (nm)	15–20	15–20
Powder density (g cm^{-3})	3.05	2.96
BET surface area (m^2 g^{-1})	45.0	75.0
Particle size from surface area (nm)	38	23
Average agglomerate size (μm)	1.78	2.54
Saturation magnetic moment (e.m.u. g^{-1})	45.1	64.3
Coercivity (Oe)	100	180
Particle size from TEM (μm)	0.20	0.15
Mol gases evolved mol^{-1} Fe$_2$O$_3$	20.0	32.0
Initiation temperature (°C)	350 ± 10	250 ± 10
Time taken to prepare γ-Fe$_2$O$_3$	5 min	2 weeks to crystallize the precursor

6.3 SPINEL FERRITES (MFe$_2$O$_4$)

Nanocrystalline spinel ferrites (MFe$_2$O$_4$) have been prepared by combustion of aqueous solutions containing stoichiometric amounts of metal nitrates and ODH redox mixtures.[3–5] In a typical experiment 1.79 g Ni(NO$_3$)$_2$ · 6H$_2$O and 5.00 g Fe(NO$_3$)$_3$ · 9H$_2$O are dissolved in a minimum amount of water (15 ml) contained in a cylindrical Pyrex dish of approximately 300 cm^3 capacity to which 2.92 g ODH is added. The mixture when placed into a muffle furnace maintained at 350°C, it boils, froths, and ignites to burn without flame yielding a voluminous dark brown powder of NiFe$_2$O$_4$ (Fig. 6.3).

Assuming complete combustion the theoretical equation for the reaction is given below:

$$M(NO_3)_2(aq) + 2Fe(NO_3)_3(aq) + 4\underset{ODH}{C_2H_6N_4O_2}\,(aq)$$

$$\longrightarrow MFe_2O_4(aq) + 8CO_2(g) + 12N_2(g) + 12H_2O$$

(32 mol of gases/mol of MFe$_2$O$_4$)

M = Mg, Mn, Co, Ni, Cu, Zn, and Cd   (3)

Fig. 6.3.   Combustion synthesized NiFe$_2$O$_4$ foam.

The composition of the redox mixtures used for the preparation of various ferrites is given in Table 6.3.

Table 6.3.   Compositions of the redox mixtures for the preparation of MFe$_2$O$_4$.

Composition of the redox mixtures	MFe$_2$O$_4$
Mg(NO$_3$)$_2$ · 6H$_2$O (1.59 g) + A + B	MgFe$_2$O$_4$
Mn(NO$_3$)$_2$ · 6H$_2$O (1.58 g) + A + B	MnFe$_2$O$_4$
Co(NO$_3$)$_2$ · 6H$_2$O (1.80 g) + A + B	CoFe$_2$O$_4$
Ni(NO$_3$)$_2$ · 6H$_2$O (1.79 g) + A + B	NiFe$_2$O$_4$
Cu(NO$_3$)$_2$ · 6H$_2$O (1.83 g) + A + B	CuFe$_2$O$_4$
Zn(NO$_3$)$_2$ · 6H$_2$O (1.84 g) + A + B	ZnFe$_2$O$_4$
Cd(NO$_3$)$_2$ · 4H$_2$O (1.90 g) + A + B	CdFe$_2$O$_4$

A = Fe(NO$_3$)$_3$ · 9H$_2$O (5.0 g), B = ODH (2.92 g).

Formation of spinel ferrites is confirmed by their characteristic powder X-ray diffraction pattern (Fig. 6.4). Lattice parameters of these ferrites calculated from powder XRDs (Table 6.4) are in good agreement with those reported in the literature.[7] The XRD patterns show considerable line broadening,

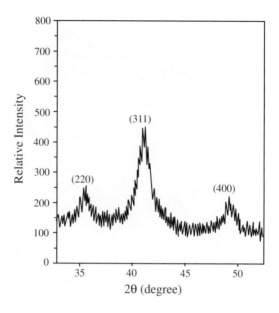

Fig. 6.4.   XRD pattern of $CoFe_2O_4$.

indicating the fine-particle nature of the ferrites. The particulate properties, e.g., density, surface area, and particle size of the green ferrites are summarized in Table 6.4. These are comparable with those prepared by single source combustible solid precursors. The green densities of the ferrites obtained by ODH processes are 65–70% of the theoretical values.

Table 6.4.   Particulate and magnetic properties of spinel ferrites.

Ferrite	Lattice constant a (nm)	Crystallite size (nm)	Powder density $(g\,cm^{-3})$	Surface area $(m^2 g^{-1})$	Particle size[a] (nm)	Saturation moment $(emu\,g^{-1})$
$MgFe_2O_4$	0.835	18	3.14 (4.12)	120 (114)	16 (13)	25.2
$MnFe_2O_4$	0.851	26	3.15 (3.27)	45 (140)	42 (14)	78.0
$CoFe_2O_4$	0.838	9	3.65 (3.65)	97 (116)	17 (15)	79.0
$NiFe_2O_4$	0.833	10	3.56 (3.57)	98 (26)	17 (68)	48.0
$CuFe_2O_4$	0.838	17	3.43	59	30	48.0
$ZnFe_2O_4$	0.843	11	3.43 (3.44)	68 (108)	26 (17)	—
$CdFe_2O_4$	0.869	—	2.20 (3.4)	40 (93)	60 (20)	—

[a] From surface area.
Values in parentheses are from precursor method.

These nanosize ferrites could be sintered to 93–98% theoretical density by pelletizing and sintering at 1050°C for 12 h. As an example, ZnFe$_2$O$_4$ achieved 98% of the theoretical density. SEM of the sintered ZnFe$_2$O$_4$ shows uniform grain sizes of 2–3 μm (Fig. 6.5).

4 μm

Fig. 6.5.   SEM of ZnFe$_2$O$_4$ sintered at 1050°C for 1 h.

6.4 MIXED METAL FERRITES

Li–Zn, Mg–Zn, and Ni–Zn ferrites are prepared by the solution combustion method and their preparation and properties are investigated.

6.4.1 *Li–Zn Ferrites*

Nanosize Li$_{0.5}$Zn$_x$Fe$_{2.5-x}$O$_4$ (where $x = 0$–1.0) powders are prepared by rapidly heating an aqueous solution containing stoichiometric amounts of the corresponding metal nitrates, iron (III) nitrate, and oxalyl dihydrazide (ODH) at 350°C.[8] The fuel taken for univalent, divalent, and trivalent metal nitrates are in the mole ratio of 1:0.5, 1:1, and 1:1.5, respectively.

When a redox mixture containing nitrates of lithium, zinc, iron, and fuel ODH is heated rapidly, it foams, froths, and deflagrates to yield voluminous ferrites that occupy the entire volume of the reaction vessel. The theoretical equations for the formation of mixed Li–Zn ferrites can be written as follows:

$$x\text{LiNO}_3(\text{aq}) + (1-x)\text{Zn(NO}_3)_2(\text{aq})$$
$$+ 2\text{Fe(NO}_3)_3(\text{aq}) + 4\text{C}_2\text{H}_6\text{N}_4\text{O}_2\ (\text{aq})$$
$$\overset{\text{ODH}}{\longrightarrow}\ \text{Li}_x\text{Zn}_{1-x}\text{Fe}_2\text{O}_4(\text{aq}) + 8\text{CO}_2(\text{g}) + 12\text{N}_2(\text{g}) + 12\text{H}_2\text{O}(\text{g})$$
$$(32\ \text{mol of gases/mol of Li}_x\text{Zn}_{1-x}\text{Fe}_2\text{O}_4) \qquad (4)$$

Composition of redox mixtures for the preparation of Li–Zn ferrites and their particulate and magnetic properties are summarized in Table 6.5.

Formation of a single spinel phase was confirmed by its characteristic X-ray powder pattern as shown in Fig. 6.6(a).

The green densities of the ferrites are in the range 3.14–3.43 g cm^{-3} (65–70% of the theoretical value) while their surface areas are in the range 50–70 m^2 g^{-1}. The lithium content of the green ferrites is estimated using atomic absorption spectroscopy (AAS). The observed values of Li are in agreement with the formulae since there is no loss of Li during combustion of the redox mixture.

The green powders (density = 45% of the theoretical value) are pressed at 6000 kg cm^{-2} and sintered in air for 1–3 h in the temperature range

Table 6.5.   Compositions of redox mixtures and particulate and magnetic properties of Li–Zn Ferrites.

Composition of the redox mixture[a]	Ferrite	Powder density ($g\,cm^{-3}$)	Surface area ($m^2\,g^{-1}$)	Particle size[b] (nm)	Saturation magnetization ($emu\,g^{-1}$)	Remnant moment ($emu\,g^{-1}$)
Fe (10.0)	$Li_{0.5}Fe_{2.5}O_4$	3.14	68	28	60.8	30.6
Fe (9.20), Zn (0.73)	$Li_{0.5}Zn_{0.2}Fe_{2.3}O_4$	3.22	72	26	63.2	31.4
Fe (8.40), Zn (1.47)	$Li_{0.5}Zn_{0.4}Fe_{2.1}O_4$	3.29	58	31	66.3	29.2
Fe (8.00), Zn (1.84)	$Li_{0.5}Zn_{0.5}Fe_{2.0}O_4$	3.40	65	27	72.1	34.6
Fe (7.60), Zn (2.20)	$Li_{0.5}Zn_{0.6}Fe_{1.9}O_4$	3.36	61	300	48.6	23.2
Fe (6.80), Zn (2.94)	$Li_{0.5}Zn_{0.8}Fe_{1.7}O_4$	3.43	53	33	24.2	12.7
Fe (10.0), Zn (3.68)	$ZnFe_2O_4$	3.43	68	26	—	—

[a] $LiNO_3 \cdot 3H_2O$ (0.61 g) + ODH (5.85 g); Values in the parenthesis indicates the weight (g) of metal nitrates.
[b] From surface area.

of 800–1100°C to get >99% density. The SEM of fractured surface of $Li_{0.5}Zn_{0.5}Fe_{2.0}O_4$ sintered at 1000°C for 3 h is shown in Fig. 6.7(a). It can be seen that the grain growth is uniform and the grain sizes are in the range of 0.7–1 μm.

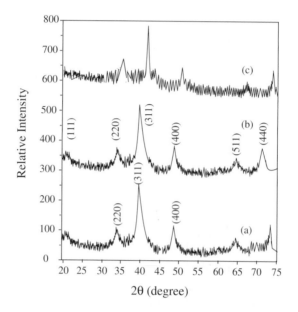

Fig. 6.6. XRD pattern of (a) $Li_{0.5}Zn_{0.5}Fe_2O_4$, (b) $Mg_{0.5}Zn_{0.5}Fe_2O_4$, and (c) $Ni_{0.5}Zn_{0.5}Fe_2O_4$.

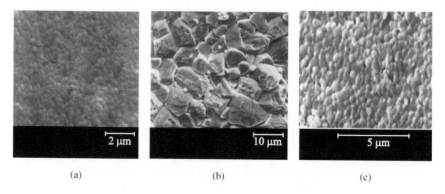

Fig. 6.7. SEM of (a) $Li_{0.5}Zn_{0.5}Fe_2O_4$ sintered at 1000°C for 1 h, (b) $Mg_{0.5}Zn_{0.5}Fe_2O_4$ sintered at 1050°C for 3 h, and (c) $Ni_{0.5}Zn_{0.5}Fe_2O_4$ sintered at 1050°C for 3 h.

6.4.2 *Mg–Zn Ferrites*

The preparation of MgZn$_{1-x}$Fe$_2$O$_4$ (where $x = 0 - 1.0$) involves the combustion of magnesium nitrate, zinc nitrate, iron (III) nitrate, ammonium nitrate, and oxalyl dihydrazide (ODH) at 350°C.[10] In a typical reaction, an aqueous solution containing 1.58 g Mg(NO$_3$)$_2 \cdot$ 6H$_2$O, 1.84 g Zn(NO$_3$)$_2 \cdot$6H$_2$O, 10.0 g Fe(NO$_3$)$_3 \cdot$9H$_2$O, and 8.76 g ODH, when rapidly heated at 350 ± 10°C, initially boils then froths and ignites to yield nanosize Mg$_{0.5}$Zn$_{0.5}$Fe$_2$O$_4$.

The theoretical equation assuming complete combustion for the reaction may be written as,

$$xMg(NO_3)_2(aq) + (1 - x)Zn(NO_3)_2(aq)$$
$$+ 2Fe(NO_3)_3(aq) + 4\underset{\text{ODH}}{C_2H_6N_4O_2}(aq)$$
$$\longrightarrow Mg_xZn_{1-x}Fe_2O_4(aq) + 8CO_2(g) + 12N_2(g) + 12H_2O(g)$$
$$(32 \text{ mol of gases/mol of } Mg_xZn_{1-x}Fe_2O_4) \qquad (5)$$

The formation of Mg$_{0.5}$Zn$_{0.5}$Fe$_2$O$_4$ seen from characteristic powder XRD pattern shows considerable line broadening, indicating the fine particle nature of Mg$_{0.5}$Zn$_{0.5}$Fe$_2$O$_4$ (Fig. 6.6b). The particulate properties of Mg$_x$Zn$_{1-x}$Fe$_2$O$_4$ are summarized in Table 6.6.

These ferrites when sintered at 900°C for 3 h show a sharp drop in magnetization at Curie temperature and reaches zero, indicating the phase purity of the Mg–Zn ferrites. The Curie temperature decreases with increase in zinc concentration.[9] This can be explained on the basis of Yaffet–Kittel type of spin arrangement.[10] The *M–H* curves of the as-prepared Mg–Zn ferrites are characterized by poor saturation demonstrating the presence of a super paramagnetic component in these ferrites. The Mg–Zn ferrites when sintered at 900°C for 3 h attain saturation because of the increase in particle size. The saturation moment as a function of zinc indicates a maximum value of 53 emu g^{-1} for Mg$_{0.5}$Zn$_{0.5}$Fe$_2$O$_4$. The remanant ratio of these sintered ferrites is around 0.65.

These ferrites could be sintered to >99% density by sintering at 1050°C for 3 h. The SEM (Fig. 6.7b) of the surface of Mg$_{0.5}$Zn$_{0.5}$Fe$_2$O$_4$ shows that the grain growth is uniform and the grain size ranges from 4 to 15 μm.

Table 6.6.   Particulate and magnetic properties of Mg–Zn ferrites.

Ferrite	Lattice constant a (nm)	Powder density (g cm^{-3})	Surface area (m^2 g^{-1})	Particle size[a] (nm)	Saturation magnetization (emu g^{-1})	Curie temperature (K)
ZnFe$_2$O$_4$	0.844	3.81	37.2	42	—	—
Mg$_{0.2}$Zn$_{0.8}$Fe$_2$O$_4$	0.843	3.62	48.2	34	28.3	—
Mg$_{0.4}$Zn$_{0.6}$Fe$_2$O$_4$	0.842	3.38	55.8	32	48.0	380
Mg$_{0.5}$Zn$_{0.5}$Fe$_2$O$_4$	0.840	3.41	64.7	27	53.0	435
Mg$_{0.6}$Zn$_{0.4}$Fe$_2$O$_4$	0.839	3.23	65.1	29	50.05	475
Mg$_{0.8}$Zn$_{0.2}$Fe$_2$O$_4$	0.838	3.19	80.1	23	33.5	550
MgFe$_2$O$_4$	0.836	3.14	110.0	17	25.2	610

[a] From surface area.

6.4.3 Ni–Zn Ferrites

Ni–Zn ferrites (Ni$_x$Zn$_{1-x}$Fe$_2$O$_4$) are prepared by the combustion of an aqueous redox mixture containing stoichiometric amounts of nickel nitrate, zinc nitrate, and ODH fuels. In a typical experiment Ni$_{0.5}$Zn$_{0.5}$Fe$_2$O$_4$ is obtained by rapidly heating a solution containing 1.801 g Ni(NO$_3$)$_2$ · 6H$_2$O, 1.84 g Zn(NO$_3$)$_2$ · 6H$_2$O, 10.00 g Fe(NO$_3$)$_3$ · 9H$_2$O, and 4.84 g ODH at 350°C.[11]

The theoretical equations assuming complete combustion for this reaction can be written as follows:

$$x\text{Ni(NO}_3)_2(\text{aq}) + (1-x)\text{Zn(NO}_3)_2(\text{aq})$$
$$+ 2\text{Fe(NO}_3)_3(\text{aq}) + 4\underset{\text{ODH}}{\text{C}_2\text{H}_6\text{N}_4\text{O}_2}(\text{aq})$$
$$\longrightarrow \text{Ni}_x\text{Zn}_{1-x}\text{Fe}_2\text{O}_4(\text{aq}) + 8\text{CO}_2(\text{g}) + 12\text{N}_2(\text{g}) + 12\text{H}_2\text{O}(\text{g})$$
$$(32 \text{ mol of gases/mol of Ni}_x\text{Zn}_{1-x}\text{Fe}_2\text{O}_4) \qquad (6)$$

The characteristic powder XRD pattern (Fig. 6.6c) shows considerable line broadening indicating the fine particle nature of Ni$_{0.5}$Zn$_{0.5}$Fe$_2$O$_4$. The particulate and magnetic properties of Ni–Zn ferrites are summarized in Table 6.7.

The magnetization curves (*M–H*) of the as-prepared ferrites at room temperature are characterized by a lower saturation value indicating the super paramagnetic nature of the ferrites. A typical *M–H* curve for Ni$_{0.5}$Zn$_{0.5}$Fe$_2$O$_4$ is given in Fig. 6.8. It can be seen that the *M–H* loop is narrow indicating high density and low porosity. The AC susceptibility measurements of Ni–Zn ferrites show super paramagnetic behavior. Although the susceptibility drops at the Curie point, it does not become zero even up to 800°C.

The Ni$_{0.5}$Zn$_{0.5}$Fe$_2$O$_4$ ferrites could be sintered to >99% density by sintering at 1050°C for 3 h. This temperature is low compared to conventional ceramic method (>1250°C).

The SEM (Fig. 6.7(c)) of the surface Ni$_{0.5}$Zn$_{0.5}$Fe$_2$O$_4$ shows that the grain growth is uniform and the grain size ranges from 0.8 to 1.0 μm.

A comparative study of nanocrystalline Ni–Zn ferrites prepared by combustible precursors (Chap. 2) as well as redox mixtures can be made. Both the methods yield magnetic materials with similar properties (Table 6.8). Nevertheless the method of making these ferrites by solution combustion of redox mixtures is fast and instantaneous and therefore more attractive.

Table 6.7.   Particulate and magnetic properties of combustion synthesized Ni–Zn ferrites.

Ferrite	Lattice constant a (nm)	Crystallite size (nm)	Powder density (g cm^{-3})	Surface area (m^2 g^{-1})	Particle size[a] (nm)	Saturation magnetization (emu g^{-1})	Curie temperature (K)
$ZnFe_2O_4$	0.844	11.3	3.43	67.5	26.0	—	—
$Ni_{0.2}Zn_{0.8}Fe_2O_4$	0.842	12.1	3.45	91.2	20.0	55.6	433
$Ni_{0.4}Zn_{0.6}Fe_2O_4$	0.840	11.7	3.47	85.6	20.0	66.0	493
$Ni_{0.5}Zn_{0.5}Fe_2O_4$	0.838	12.7	3.47	90.1	20.0	73.4	545
$Ni_{0.2}Zn_{0.8}Fe_2O_4$	0.837	12.1	3.48	85.2	21.0	62.1	628
$Ni_{0.2}Zn_{0.2}Fe_2O_4$	0.835	13.2	3.43	91.3	20.0	62.1	718
$Ni\,Fe_2O_4$	0.833	10.3	3.56	98.1	17.0	43.2	863

[a] From surface area.

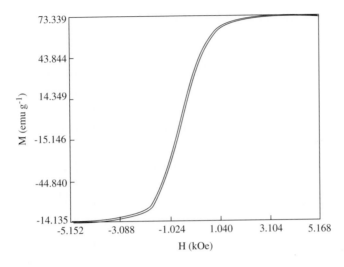

Fig. 6.8.   *M–H* curve of Ni$_{0.5}$Zn$_{0.5}$Fe$_2$O$_4$ sintered at 900°C for 3 h.

Table 6.8.   Comparison of properties of Ni–Zn ferrites prepared by solution combustion and solid combustible precursor methods.

Property	Combustion method	Precursor method
Crystallite size (X-ray line broadening)	10–15 nm	10–15 nm
BET surface area	85–95 m^2 g^{-1}	42–108 m^2 g^{-1}
Powder density (% theoretical density)	65–75%	65–75%
TEM (particle size)	60–90 nm	60–90 nm
Sinterability (>98% density)	1050°C, 3 h	1000°C, 24 h
Average grain size (SEM)	3–5 μm	1–2 μm
Initiation temperature	350 ± 10°C	250 ± 40°C
Homogeneity	100%	100%
Number of moles of gas evolved	32 mol	45 mol
Time taken to prepare Ni–Zn ferrite	Instantaneous (3 min)	1–2 weeks to prepare precursor; decomposition of precursor in <5 min

6.5 RARE EARTH ORTHOFERRITES

Rare earth orthoferrites ($LnFeO_3$) are prepared by the combustion of aqueous solutions containing corresponding rare earth nitrate, ferric nitrate, and ODH mixtures.[3] The actual weights of redox mixtures used for combustion are given in Table 6.9. Assuming complete combustion, the theoretical equation for the formation of rare earth orthoferrite may be written as follows:

$$Ln(NO_3)_2(aq) + Fe(NO_3)_3(aq) + 3\,C_2H_6N_4O_2\,(aq)$$
$$\text{ODH}$$
$$\longrightarrow LnFeO_3(s) + 6CO_2(g) + 9N_2(g) + 9H_2O(g)$$
$$(24\,\text{mol of gases/mol of }LnFeO_3)$$
$$Ln = La,\,Nd,\,Sm,\,Gd,\,Dy,\,\text{and Y} \qquad (7)$$

The as-prepared rare earth orthoferrites which are X-ray amorphous form crystalline structure on heating at 750°C for 1 h (Fig. 6.9).

The particulate properties of various orthoferrites prepared are summarized in Table 6.9.

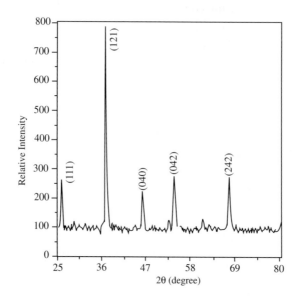

Fig. 6.9.   XRD pattern of $SmFeO_3$.

Table 6.9.   Compositions of redox mixtures and particulate properties of orthoferrites.

Composition of the redox mixtures	Orthoferrites	Powder density $(g\,cm^{-3})$	Surface area $(m^2\,g^{-1})$	Particle size[a] (nm)
Y (4.74 g) + A	YFeO$_3$	3.82	45.7	34
La (5.36 g) + A	LaFeO$_3$	4.21	22.1	65
Nd (5.43 g) + A	NdFeO$_3$	4.42	45.0	30
Sm (5.50 g) + A	SmFeO$_3$	4.52	37.3	36
Gd (5.59 g) + A	GdFeO$_3$	4.21	16.8	85
Dy (5.65 g) + A	DyFeO$_3$	4.68	29.2	44

A = Fe(NO$_3$)$_3$ · 9H$_2$O (5.0 g) + ODH (4.38 g). Values in the parenthesis are the weights of rare earth metal nitrates.
[a] From surface area.

6.6 GARNETS (Ln$_3$Fe$_5$O$_{12}$)

The rare earth iron garnets Ln$_3$Fe$_5$O$_{12}$ and Y$_3$Al$_x$Fe$_{5-x}$O$_{12}$ (where $x = 1.0$–5.0) are prepared by the combustion of aqueous solutions containing redox mixtures of corresponding metal nitrates and ODH.[12] The actual weights of redox mixtures used for combustion are given in Table 6.10. The solution undergoes combustion, to yield fine, foamy, amorphous garnets (foam density of 0.001 g cm^{-3} and tap density of 0.01 g cm^{-3}) in less than 5 min. The theoretical equation assuming complete combustion may be written as follows:

$$3Ln(NO_3)_3(aq) + 5Fe(NO_3)_3(aq) + \underset{ODH}{12C_2H_6\,N_4O_2}$$

$$\longrightarrow Ln_3Fe_5O_{12}(s) + 24CO_2(g) + 36N_2(g) + 36H_2O(g)$$

$$\text{(96 mol of gases/mol of } Ln_3Fe_5O_{12})$$

$$Ln = Sm, Gd, Dy, Eu, Tb, \text{ and } Y \qquad (8)$$

Similarly, gadolonium substituted for yttrium in Y$_3$Al$_x$Fe$_{5-x}$O$_{12}$ forms Y$_{1.5}$Gd$_{1.5}$Al$_{0.2}$Fe$_{4.8}$O$_{12}$ by the combustion of aqueous solutions containing redox mixtures of corresponding metal nitrates and ODH.

Table 6.10. Compositions of redox mixtures and particulate properties of rare earth iron garnets and yttrium aluminum garnets.

Composition of the redox mixtures	Garnets	Lattice constant a (nm)	Powder density (g cm^{-3})	Surface area (m^{2}g^{-1})	Particle size[a] (nm)
Sm (3.30 g) + A	$Sm_3Fe_5O_{12}$	1.253	4.21	22.7	63
Eu (3.31 g) + A	$Eu_3Fe_5O_{12}$	1.250	4.32	29.4	47
Dy (3.39 g) + A	$Dy_3Fe_5O_{12}$	1.241	4.17	32.1	45
Gd (3.35 g) + A	$Gd_3Fe_5O_{12}$	1.247	4.27	3.2	36
Tb (3.36 g) + A	$Tb_3Fe_5O_{12}$	1.243	4.12	27.0	54
Y (2.84 g) + A	$Y_3Fe_5O_{12}$	1.234	2.76	89.1	24
Fe (4.0 g), Al (0.93 g) + B	$Y_3AlFe_4O_{12}$	1.227	2.91	65.3	32
Fe (3.0 g), Al (1.86 g) + B	$Y_3Al_2Fe_3O_{12}$	1.221	2.95	38.9	52
Fe (2.0 g), Al (2.78 g) + B	$Y_3Al_3Fe_2O_{12}$	1.215	3.10	27.9	70
Fe (1.0 g), Al (3.71 g) + B	$Y_3Al_4FeO_{12}$	1.208	3.32	16.3	111
Y (142 g), Gd (1.68 g) + A	$Y_{1.5}Gd_{1.5}Fe_5O_{12}$	1.240	3.92	49.5	31

A = Fe(NO$_3$)$_3$ · 9H$_2$O (5.0 g) + ODH (3.5 g), B = Y(NO$_3$)$_3$ · 6H$_2$O (2.84 g). Values in the parenthesis are the weights of rare earth metal nitrates.
[a]From surface area.

As-prepared garnet powders are X-ray amorphous and crystallize to form a single-phase product on calcination at 750°C for 3 h (Fig. 6.10). As the aluminum content increases in the redox mixture, the reaction proceeds from smouldering to flaming type, yielding crystalline phase. The particulate properties of rare earth iron garnets and yttrium aluminum garnets are summarized in Table 6.10.

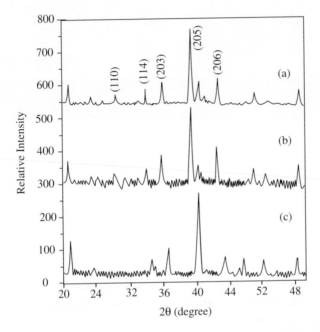

Fig. 6.10.   XRD patterns of heat treated (750°C for 3 h): (a) Y$_3$Fe$_5$O$_{12}$, (b) Sm$_3$Fe$_5$O$_{12}$, and (c) Y$_3$Al$_2$Fe$_3$O$_{12}$.

Yttrium iron garnet can be sintered to a density of more than 95% at 1200°C for 3 h. The saturation magnetization of the sintered YIG shows magnetization value of 1750 G, which is comparable with the reported value.[13] The scanning electron micrograph of the surface of sintered YIG is shown in Fig. 6.11. It shows a uniform grain size of 3–5 μm.

Fig. 6.11.   SEM of $Y_3Fe_5O_{12}$ sintered at 1200°C for 3 h

6.7 BARIUM AND STRONTIUM HEXAFERRITES

Hexaferrites ($MFe_{12}O_{19}$ where M = Sr and Ba) are prepared by the solution combustion of redox mixtures containing corresponding metal nitrates and ODH.[3] Barium hexaferrite is also prepared using other fuels like MDH and MH.[14] The theoretical equation for the formation of hexaferrites using ODH fuel may be written as follows:

$$M(NO_3)_2(aq) + 12Fe(NO_3)_3(aq) + 19 \underset{ODH}{C_2H_6N_4O_2} (aq)$$
$$\longrightarrow MFe_{12}O_{19}(s) + 38CO_2(g) + 57N_2(g) + 57H_2O(g)$$
$$(152 \text{ mol of gases/mol of } MFe_{12}O_{19})$$
$$M = Ba, Sr \quad (9)$$

The combustion product of the above mixture is X-ray amorphous. It becomes crystalline when heated at 850°C for 3 h. The particulate properties of these compounds are summarized in Table 6.11.

Table 6.11. Particulate properties of hexaferrites.

Hexafarrite	Fuel	Lattice constant (nm)	Powder density (g cm^{-3})	Surface area (m^2 g^{-1})	Particle size[a] (nm)
BaFe$_{12}$O$_{19}$	ODH	a = 0.593, b = c = 2.327	2.9	62.0	32.0
	MDH	a = 0.591, b = c = 2.326	3.0	42.0	47.0
	MH	a = 0.592, b = c = 2.324	3.3	35.0	51.0
SrFe$_{12}$O$_{19}$	ODH	a = 0.588, b = c = 2.322	3.6	24.2	70.0

[a] From surface area.

The magnetic properties of combustion synthesized hexaferrites as a function of sintering temperature[15] is illustrated in Table 6.12

The as-formed samples show lower M_s and H_c values, which may be due to the small particles approaching superparamagnetic size. As sintering

Table 6.12. Particulate and magnetic properties of (Ba/Sr)Fe$_{12}$O$_{19}$.

Sintering temperature (°C)	Crystallite size (nm)	M_s (emu g^{-1})	M_r (emu g^{-1})	M_r/M_s	H_c (Oe)
BaFe$_{12}$O$_{19}$					
As-prepared	—	26.43	14.66	0.55	593
800	47	84.00	41.33	0.49	4697
1000	62	83.00	40.67	0.49	4620
1100	80	96.00	55.46	0.59	3670
1250	—	95.58	54.57	0.57	1306
SrFe$_{12}$O$_{19}$					
As-prepared	—	24.38	13.34	0.55	352
800	35	83.33	47.58	0.57	5525
1000	65	90.90	45.90	0.50	6187
1100	100	104.90	64.23	0.61	4909
1250	—	115.00	68.88	0.59	4203
1350	—	111.25	35.24	0.32	926

temperature is increased, higher values of saturation magnetization and coercive force are attained. However, above 1000°C, H_c gradually decreases for $SrFe_{12}O_{19}$ presumably because of grain growth. This has been confirmed by SEM (Fig. 6.12).

Fig. 6.12. SEM of $SrFe_{12}O_{19}$ sintered (a) as-compacted, (b) 600°C, (c) 900°C, and (d) 1250°C (2 h).

The magnetic properties of combustion synthesized $BaFe_{12}O_{19}$ has been compared with those reported in literature (Table 6.13). Combustion derived nano-$BaFe_{12}O_{19}$ shows higher M_s and M_r values compared to other reported values.[16]

Table 6.13.   Particulate and magnetic properties of BaFe$_{12}$O$_{19}$ obtained by different routes.

Preparation method	Particle shape	Particle size (μm)	M_s (emu g^{-1})	M_r (emu g^{-1})	H_c (Oe)
Ceramic	Platelet	1–4	78.2	39.4	2178
Nitrate decomposition	Platelet	0.15–0.6	70.2	34.3	4320
Liquid mix technique	Spherical	0.1–0.15	74.9	36.2	5550
Pyrolysis citrate	Spherical	0.03–0.05	73.0	35.5	5340
Pyrolysis nitrate	Spherical	0.03–0.04	70.7	34.5	4867
Solution combustion	Spherical	0.03–0.04	84.0	41.3	4697

6.8 CONCLUDING REMARKS

Nanocrystalline magnetic oxides (γ-Fe$_2$O$_3$, spinel ferrites, mixed ferrites, rare earth orthoferrites, garnets, and hexaferrites) are prepared by solution combustion method using corresponding metal nitrates and ODH. Orthoferrites, garnets, and hexaferrites which cannot be prepared by the combustible solid precursors are readily prepared by solution combustion synthesis. Being nanocrystalline in nature, these magnetic materials exhibit superparamagnetism and find applications for making thin films (sensor applications), ferrofluids, stealth bombers, cancer detection, and remediation, etc.

References

1. Sugimoto M, The past, present and future of ferrites, *J Am Ceram Soc* **82**: 269–280, 1999.
2. Gould P, Nanomagnetism shows in vivo potential, *Nanotoday* **1**: 34–39, 2006.
3. Suresh K, Patil KC, A recipe for an instant synthesis of fine particle oxide materials, in Rao KJ (ed.), *Perspectives in Solid State Chemistry*, Narosa Publishing House, New Delhi, pp. 376–388, 1995.
4. Arul Dhas N, Muthuraman M, Ekambaram S, Patil KC, Synthesis and properties of fine particle cadmium ferrite (CdFe$_2$O$_4$), *Int. J Self-Propagating High-Temp Synth* **3**: 39–50, 1994.
5. Suresh K, Kumar NRS, Patil KC, A novel combustion synthesis of spinel ferrites, orthoferrites and garnets, *Adv Mater* **3**: 148–150, 1991.

6. Suresh K, Patil KC, A combustion process for the instant synthesis of γ-Fe_2O_3, *J Mater Sci Lett* **12**: 572–574, 1993.

7. Powder Diffraction File, Joint Committee on Diffraction Standards, Pennsylvania, 1988.

8. Suresh K, Patil KC, Combustion synthesis and properties of fine particle Li–Zn ferrites, *J Mater Sci Lett* **14**: 1074–1077, 1995.

9. Suresh K, Patil KC, Combustion synthesis and properties of fine particle Mg–Zn ferrites, *J Mater Sci Lett* **13**: 1712–1714, 1994.

10. Satyamurthy NS, Natera MG, Yousef SI, Begam RJ, Srivatava CM, Yafet–Kittel angles in zinc–nickel ferrites, *Phys Rev* **181**: 969–977, 1969.

11. Suresh K, Patil KC, Preparation and properties of fine particle nickel–zinc ferrites: A comparative study of combustion and precursor method, *J Solid State Chem* **99**: 12–17, 1992.

12. Suresh K, Patil KC, Combustion synthesis and properties of $Ln_3Fe_5O_{12}$ and yttrium aluminium garnets, *J Alloys Compd* **209**: 203–206, 1994.

13. Multani M, Nandikar NG, Venkataramani M, Raghupathy V, Pansrare AK, Gurjar A, Hot-sprayed yttrium iron garnet, *Mater Res Bull* **14**: 1251–1258, 1979.

14. Aruna ST, Patil KC, Combustion synthesis and magnetic properties of nanosize barium hexaferrite, *Trans Ind Ceram Soc* **55**: 147–150, 1996.

15. Aruna ST, *Combustion synthesis and properties of nanomaterials: Synthesis and properties of solid oxide fuel cell materials*, PhD Thesis, Indian Institute of Science, Bangalore, 1998.

16. Vallet Regi M, Preparative strategies of controlling structure and morphology of metal oxides, in Rao KJ (ed.), *Perspectives in Solid state Chemistry*, Narosa Publishing House, New Delhi, pp. 37–65, 1995.

17. Patil KC, Mimani T, Preparation and properties of nano crystalline magnetic oxides, *Magn Soc Ind Bull* **22**: 21–26, 2000.

Chapter 7

Nano-Titania and Titanates

7.1 INTRODUCTION

Titania exists naturally in three important crystallographic forms: anatase, brookite, and rutile. Brookite is a naturally occurring phase and is extremely difficult to synthesize. The structure of rutile and anatase can be described in terms of chains of TiO_6 octahedra (Fig. 7.1). The two crystal structures differ by the distortion inside each octahedron and their assemblage. Each Ti(IV) ion at the center of octahedron is surrounded by six O^{2-} ions (Fig. 7.2). Anatase is a thermodynamically unstable phase, with the phase transition to rutile becoming detectable at 700°C.

The interest in titania and titanates is due to their photocatalytic[1-4] and dielectric properties. Titanates are of use due to their high sinteractivity and reactivity to assimilate radioactive nuclear waste. Some of the important properties and applications of titania and titanates are summarized in Table 7.1.

The preparation and properties of perovskite titanates ($MTiO_3$) and pyrochlores ($Ln_2Ti_2O_7$) are discussed in Chaps. 2 and 9, respectively. In this chapter, the methodology for synthesis of nano-titania, metal-ion-substituted titania, and titanates (synroc materials) used for nuclear waste immobilization by solution combustion is presented. The particulate nature, structure, morphology, and photocatalytic properties of combustion derived nano-titania and metal-ion-substituted titania have been investigated. The potential application of Pd-doped TiO_2 as a three-way catalyst for auto exhaust has been explored.

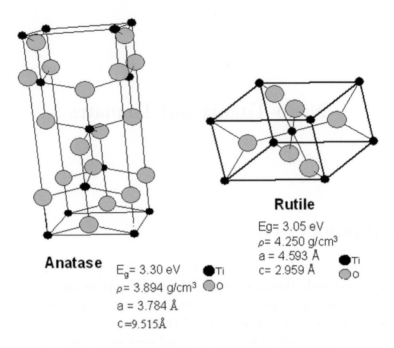

Rutile

Eg= 3.05 eV
ρ= 4.250 g/cm³
a = 4.593 Å
c= 2.959 Å

●Ti
◐O

Anatase E_g= 3.30 eV   ●Ti
ρ= 3.894 g/cm³ ◐O
a = 3.784 Å
c=9.515Å

Fig. 7.1.   The crystal structures of anatase and rutile TiO_2.

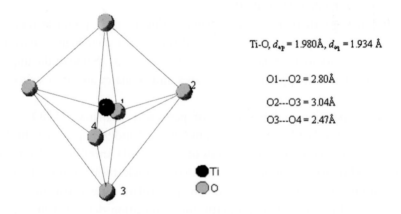

Ti-O, d_{ap} = 1.980Å, d_{eq} = 1.934 Å

O1---O2 = 2.80Å

O2---O3 = 3.04Å

O3---O4 = 2.47Å

●Ti
◯O

Fig. 7.2.   Octahedron structure of TiO_2.

Table 7.1. Properties and applications of titania and titanates.

Compound	Applications	Applications
TiO_2	Optical property (high refractive index $n = 2.4$)	White pigment, reflective optical coating for dielectric mirrors and gemstones, paints, coatings, plastics, papers, inks, foods, medicines (i.e., pills and tablets), toothpastes, cosmetic, skin care, ceramic glazes, sunscreen
	Semiconducting	Photocatalyst (organic matter), dye-sensitized solar cells
	Hydrophobic	Antifogging coatings
	Resistance	Resistance-type lambda probes (oxygen sensor)
Nano-TiO_2	Optical property	Monochromatic screen which is energetically self-sufficient
Size 20–50 nm	Transparent to visible light	Invisible sunscreens
Size 50–100 nm	Scattering	White paint
Size >100 nm	Semiconducting	Photocatalyst for volatile organics as well as airborne pollutants, H_2S and pollutants, self-cleaning windows
M^0/TiO_2 Metal-ion-substituted TiO_2	Photocatalysis Catalysis	Oxidation of organic pollutants (dyes, oil spills) TWC for automobile exhaust
$MTiO_3$ (M = Ca, Sr, Ba, Pb), PZT	Dielectric/pyroelectric/piezoelectric	Capacitors, acoustic transducer, sensors, pyrodetectors, photo anodes for solar energy conversion
$Ln_2Ti_2O_7$	Relaxor ferroelectrics	High-frequency devices, microwave resonator
Synroc materials ($CaTiO_3$, $CaZrTi_2O_7$, $CaTiSiO_5$)	Reactivity to assimilate radioactive waste materials	Nuclear waste immobilization

7.2 NANO-TIO$_2$ (ANATASE)

It is generally accepted that the anatase phase is more photo active than the rutile phase. This can be understood by viewing the molecular orbital bonding diagram of TiO$_2$ (Fig. 7.3). The valence band (VB) and conduction band (CB) consist of both Ti "3d" and O "2p" orbitals. As the Ti "3d" orbital splits into t_{2g} and e_g states, the CB is divided into lower and upper parts. The VB and the upper CB are composed of O "2p" and Ti "e_g" states, while the lower CB consists of the O "2p" and Ti "t_{2g}" states. The VB can be decomposed into three parts: bonding of the O "pσ" and Ti "e_g" states in the lower energy region; the π bonding of the O "pπ" and Ti "e_g" states in the middle energy region; and O "pπ" states in the higher energy region.

The bottom of the lower CB consisting of the Ti "d$_{xy}$" orbital contributes to the metal–metal interactions due to the σ bonding of the Ti "t_{2g}" – Ti "t_{2g}" states. At the top of the lower CB, the rest of the Ti "t_{2g}" states are antibonding with the O "pπ" states. The upper CB consists of the σ antibonding between the O "pσ" and Ti "e_g" states. Thus, the CB has strong metal d-orbital character, originating from the antibonding metal–oxygen interaction.

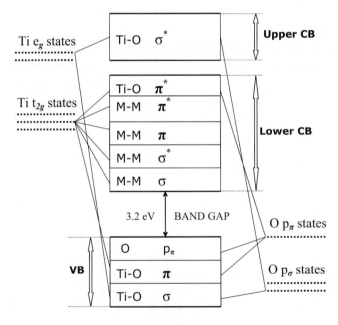

Fig. 7.3.   Molecular orbital bonding diagram of TiO$_2$.

This difference between the two forms of titania is usually explained by the fact that conduction band of anatase is about 0.3 eV more positive than rutile phase. Consequently, the excess electrons in the conduction band of anatase possesses more driving force for the reduction of reactants than that of rutile. Additionally, differences in the surface hydroxylation of these solids may also contribute to variation in their activities. The lesser activity of rutile is also attributed to its lower capacity to adsorb O_2, which is necessary to prevent electron–hole recombination. Therefore in both photocatalysis and in photochemical applications anatase TiO_2 is used.

7.2.1 Synthesis and Properties of Nano-TiO₂ (Anatase)

Nanosize titania (anatase) is prepared by the combustion of aqueous solution containing stoichiometric amounts of $TiO(NO_3)_2$ and fuels like tetraformal trisazine (TFTA), glycine (GLY), hexamethylenetetramine (HMT), oxalyldihydrazide (ODH), and carbohydrazide (CH).[5-7] The stoichiometry of the metal nitrate and fuel mixture is calculated based on the total oxidizing and reducing valences of the oxidizer and the fuel. In a typical experiment, a Pyrex dish (300 cm^3) containing an aqueous redox mixture of 2 g titanyl nitrate, 0.89 g glycine in 15 ml of water is introduced into a muffle furnace preheated to 350°C. The solution undergoes dehydration, gives a spark, which appears at one corner and spreads throughout the mass. The combustion reaction is of smoldering type without the appearance of flame. Theoretical equations for the formation of TiO_2 by the combustion of the titanyl nitrate with various fuels can be represented as

$$TiO(NO_3)_2(aq) + \underset{CH}{CH_6N_4O(aq)}$$
$$\longrightarrow \underset{C}{4TiO_2(s)} + 14N_2(g) + 5CO_2(g) + 15H_2O(g)$$
$$(8.5 \text{ mol of gases/mol of TiO}_2) \qquad (1)$$

$$9TiO(NO_3)_2(aq) + \underset{GLY}{10C_2H_5O_2N(aq)}$$
$$\longrightarrow \underset{G}{9TiO_2(s)} + 14N_2(g) + 20CO_2(g) + 25H_2O(g)$$
$$(\sim 6.5 \text{ mol of gases/mol of TiO}_2) \qquad (2)$$

$$18TiO(NO_3)_2(aq) + 5C_6H_{12}N_4(aq)$$
$$\underset{HMT}{}$$
$$\longrightarrow 18TiO_2(s) + 28N_2(g) + 30CO_2(g) + 30H_2O(g)$$
$$\underset{H}{}$$

$$(\sim4.8 \text{ mol of gases/mol of TiO}_2) \qquad (3)$$

$$14TiO(NO_3)_2(aq) + 5C_4H_{16}N_6O_2(aq)$$
$$\underset{TFTA}{}$$
$$\longrightarrow 14TiO_2\,(s) + 29N_2(g) + 20CO_2(g) + 40H_2O(g)$$
$$\underset{T}{}$$

$$(\sim6.3 \text{ mol of gases/mol of TiO}_2) \qquad (4)$$

$$TiO(NO_3)_2(aq) + C_2H_6N_4O_2(aq)$$
$$\underset{ODH}{}$$
$$\longrightarrow TiO_2(s) + 3N_2(g) + 2CO_2(g) + 3H_2O(g)$$
$$\underset{O}{}$$

$$(8 \text{ mol of gases/mol of TiO}_2) \qquad (5)$$

Titania obtained by the combustion of redox mixtures of various fuels are hereafter referred to $TiO_2(C)$, $TiO_2(G)$, etc. after the fuel used. The yellow color observed for titania derived from ODH and glycine has been attributed to the partial substitution of oxide ion in TiO_2 by carbide and nitride ions.[6]

Anatase TiO_2 is characterized by XRD pattern, Raman spectroscopy, UV spectroscopy, TEM, and surface area measurements. The particulate properties of TiO_2 obtained by the combustion of redox mixtures are summarized in Table 7.2.

The powder XRD patterns (Fig. 7.4) of titania confirm the formation of anatase phase and show considerable line broadening indicating nanostructured titania. The crystallite sizes of combustion synthesized titania calculated from Debye–Scherrer formula are in the range of 6–9 nm whereas commercial Degussa P25 TiO_2 has a crystallite size of 30–35 nm. Rietveld refined X-ray diffraction pattern of TiO_2 prepared from glycine is given in Fig. 7.5. There is a good agreement between calculated and observed patterns. It may be noted that background of the X-ray pattern is flat indicating that combustion synthesized TiO_2 is crystalline. The R_{Bragg}, R_F, and R_P, values are 2.46, 1.60, and 4.87%, respectively. The total oxygen in all the TiO_2 samples is 2. The refinements of all combustion-derived TiO_2 samples are in good agreement with the calculated pattern with no significant variation in the lattice parameters and oxygen content.

Table 7.2.   Particulate properties of anatase titania.

Redox mixture	Product	Powder density $(g\,cm^{-3})$	Surface area $(m^2\,g^{-1})$	Particle size[a] (nm)	Crystallite size (nm)	Anatase lattice parameters (nm)
A + CH	TiO_2 (C)	3.30	100	18	6	$a = 0.37639$ $c = 0.95147$
A + GLY	TiO_2 (G) yellow	3.12	156	12	6	$a = 0.37865$ $c = 0.95091$
A + HMT	TiO_2 (H) white	3.10	164	11	7	$a = 0.37865$ $c = 0.95091$
A + ODH	TiO_2 (O) pale yellow	3.20	90	20	9	$a = 0.37707$ $c = 0.95501$
A + TFTA	TiO_2 (T)	3.35	78	23	9	$a = 0.37855$ $c = 0.9515$

$A = TiO(NO_3)_2$.
[a] From surface area.

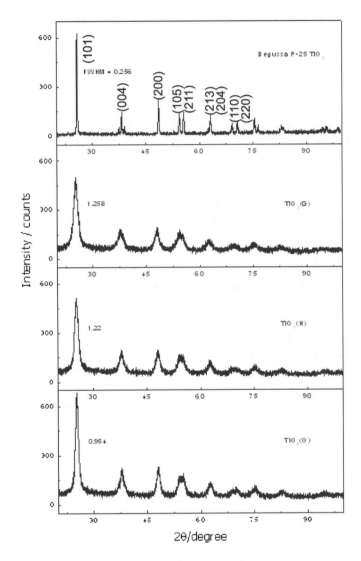

Fig. 7.4. XRD patterns of combustion synthesized titania and commercial titania.

The phase purity of combustion-synthesized titania powders is also confirmed by Raman spectra, which substantiates the presence of only anatase phase (Fig. 7.6).

The anatase phase stability of combustion-synthesized titania is investigated by calcination at various temperatures (500°C, 575°C, 650°C, 850°C

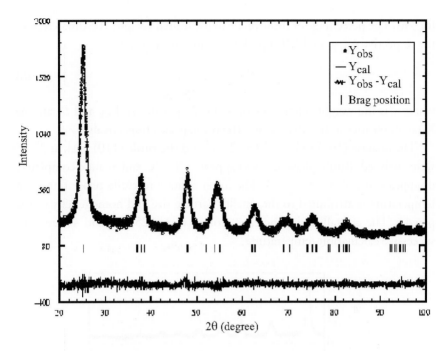

Fig. 7.5.   Rietveld refined X-ray diffraction pattern of TiO_2 prepared from glycine.

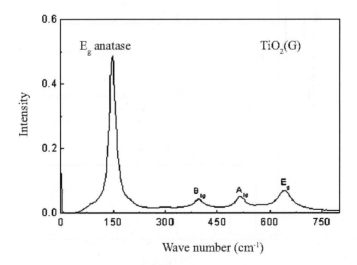

Fig. 7.6.   Raman spectrum of $TiO_2(G)$.

for 5 h). The percentage of rutile phase in the heated samples is estimated from the respective integrated XRD peak intensities using Eq. (6):

$$x = (1 + 0.8I_A/I_R)^{-1} \tag{6}$$

where x is the weight fraction of rutile in the powder and I_A and I_R are the X-ray intensities of the anatase and the rutile peaks, respectively.

The anatase (101) peak at $2\theta = 25.4°$ and the rutile (110) peak at $27.5°$ were analyzed. Rutile phase starts to appear at $575°C$ and anatase completely disappears at $850°C$ (Fig. 7.7). The lower anatase to rutile phase transition temperature is attributed to the smaller particle size and homogeneity of the powder. The crystallite size increases from 6 to 14, 25, 33, and 45 nm with successive calcination temperatures.

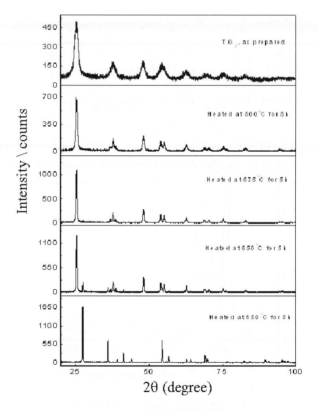

Fig. 7.7.   XRD patterns of as-prepared TiO_2 and TiO_2 heated at different temperatures.

The nanosize nature of combustion-synthesized titania powders is also substantiated from TEM studies. The bright-field TEM images of $TiO_2(G)$, $TiO_2(H)$, and $TiO_2(O)$, and Degussa P25 TiO_2 are shown in Figs. 7.8(a) to 7.8(d). Dark-field imaging is used to identify the individual particles. The particle sizes determined from TEM for combustion-synthesized titania are in the range of 6–13 nm whereas Degussa P25 TiO_2 shows larger particles in the range of 100–150 nm. Analysis of lattice fringes of individual small crystallites in the high-resolution TEM image of $TiO_2(G)$ (Fig. 7.9a) shows a lattice spacing of 3.5 ± 0.1 Å, which is in good agreement with the anatase (101) lattice spacing of 3.52 Å. The electron diffraction studies indicate that combustion-synthesized titania powders are highly crystalline (absence of diffuse halo) and the pattern can be indexed to anatase phase only (Fig. 7.9b). The d-spacing calculated from the ED pattern also matches with the values obtained from XRD patterns.

7.3 PHOTOCATALYTIC PROPERTIES OF NANO-TiO_2

In recent years, the wide spread presence of chemicals such as heavy metals, herbicides, pesticides, aliphatic and aromatic detergents, arsenic compounds, solvents, degreasing agents, volatile organics, and chlorophenols pose a serious threat to the environment. When such chemicals contaminate water sources, they become really hazardous. For instance, waste waters produced from textile and dyestuff industrial processes contain large quantities of azo dyes. It is estimated that 15% of the total dye is lost during dying process and released in waste waters. Oxidation of these organic pollutants at the surface of TiO_2 catalyst is an important photocatalysis application.

Heterogeneous photocatalysis is a process in which a combination of photochemistry and catalysis are operating together. It implies that both light and catalyst are necessary to bring out the chemical reaction. UV light illumination over a semiconductor like TiO_2 produces electrons and holes. The valence band holes are powerful oxidants ($+1$ to $+3.5$ V versus NHE depending on the semiconductor and pH), while the conduction band electrons are good reductants ($+0.5$ to -1.5 V versus NHE). In 1977, Frank and Bard[2,3] examined the possibilities of decomposing cyanide in water by titania. Since then there is an increasing interest in semiconductor-mediated photo-oxidative

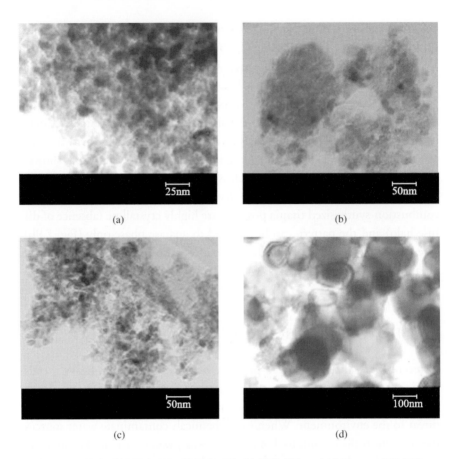

Fig. 7.8.   TEM of (a) TiO$_2$(G), (b) TiO$_2$(H), (c) TiO$_2$(O), and (d) Degussa P25 TiO$_2$.

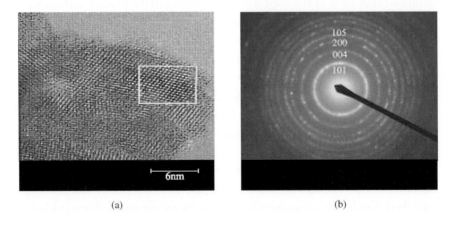

Fig. 7.9.   (a) HRTEM image of TiO$_2$ (G) and (b) electron diffraction pattern of TiO$_2$ (G).

degradation by titania for environmental applications. Beck and Siegel[4] have studied the decomposition of H_2S as a function of surface area (particle size) of titania to demonstrate the improved catalytic property of titania.

The elimination of organic pollutants is achieved by advanced oxidation technologies (AOT) which are based on generation of highly reactive intermediates (e.g., hydroxyl radicals, $O_2^{\bullet-}$), that initiate sequence of reactions resulting in the destruction of such pollutants. Chemical reactions under O_3/UV, H_2O_2/UV, and semiconductor/air/UV are the main advanced photocatalytic oxidation techniques that show high promise in this area. Oxidation potentials of common oxidants are listed in Table 7.3, which shows hydroxyl radicals to be the most powerful oxidizing species.

Table 7.3.   Oxidation potentials of some oxidants.

Species	Oxidation potential (V)
Hydroxyl radicals	2.80
Atomic oxygen	2.42
Ozone	2.07
Hydrogen peroxide	1.78
Per hydroxyl radicals	1.70
Permanganate	1.68
Iodine	0.54

Irradiation of TiO_2 with light energy equal to or greater than the band gap promotes an electron from the valence band to the conduction band, creating an electronic vacancy or "hole" (h^+) at the valence band edge. This is a rapid process occurring in femto seconds (Fig. 7.10).

The mechanism of photochemical oxidation reactions of titania can be written as follows:

Primary Process. Charge-carrier generation:

$$TiO_2 + h\upsilon \longrightarrow h_{VB}^+ + e_{CB}^- \tag{7}$$

Once excitation occurs across the band gap there is sufficient life time, in the nanosecond regime, for the created electron–hole pair to undergo charge transfer process. The conduction band electrons and valence band holes can

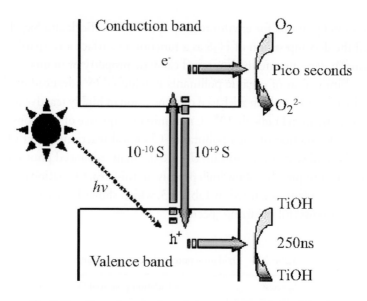

Fig. 7.10.   General mechanism of photocatalytic activity of titania.

be trapped at the titania surface. Trapped holes react with water or hydroxide ions adsorbed on the surface to form hydroxyl radicals:

$$h_{VB}^{+} + 2OH_{(ads)}^{-} \longrightarrow OH^{-} + OH^{\bullet} \tag{8}$$

$$H_2O_{(ads)} + h_{VB}^{+} \longrightarrow H^{+} + OH^{\bullet}$$
$$\updownarrow \tag{9}$$
$$(H^{+} + OH^{-})$$

On the other hand, the trapped electrons can be scavenged by reducible species, such as O_2:

$$e_{CB}^{-} + O_2 \longrightarrow O_2^{\bullet -} \tag{10}$$

$$2O_2^{\bullet -} + 2H_2O \longrightarrow 2OH^{\bullet} + 2OH^{-} + O_2 \tag{11}$$

Finally, oxidation of the organic reactant occurs via successive attacks by $OH^{\bullet}$ radicals:

$$R + 4OH^{\bullet} \longrightarrow R'^{\bullet} + 2H_2O + O_2 \tag{12}$$

In this way, the organic radical undergoes total oxidation to carbon dioxide and water.

It is well known that surface hydroxyl groups present on titania play an important role in the photocatalytic degradation of organic water contaminants. Higher the number of surface hydroxyl groups higher is the photocatalytic activity. To confirm the presence of surface hydroxyl groups, thermogravimetric analysis is carried out for combustion-synthesized and commercial titania in oxygen ($150\, cc\, min^{-1}$) at the heating rate of $5°C\, min^{-1}$ (Fig. 7.11). The total weight loss observed in the TGA experiments are higher for solution combustion-synthesized titania (6–15%) compared to 1.7% observed for Degussa P-25 titania. Among the different samples of TiO_2 powders, TiO_2 prepared from glycine has highest surface hydroxylation.

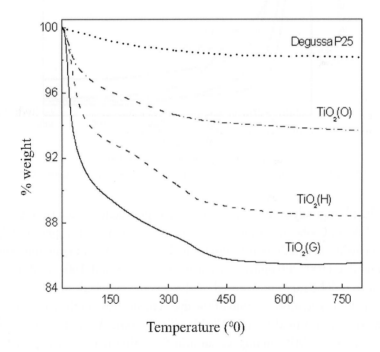

Fig. 7.11.   TGA of combustion synthesized and Degussa P25 TiO_2.

The band gap energy values (E_g) of the combustion-synthesized titania powders calculated from UV–Visible absorption spectra (Fig. 7.12) are found to be much lower (2.21–2.98 eV) than commercial TiO_2 (3.3 eV). This indicates more effective photocatalytic behavior of combustion-synthesized titania. Also, the photoluminescence (PL) emission spectra (Fig. 7.13) of combustion-synthesized TiO_2 show higher intensity when compared to commercial TiO_2 nanoparticles. This difference in intensity is attributed to the large amounts of adsorbed surface water on the combustion synthesized titania as compared to commercial titania.

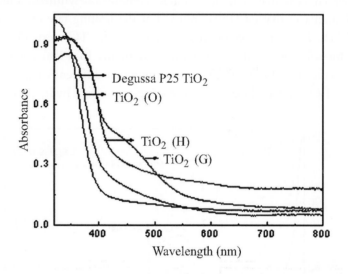

Fig. 7.12.   UV–Vis absorption spectra of combustion synthesized and Degussa P25 TiO_2.

The photocatalytic activity of titania is experimentally determined using both UV and solar radiations. A schematic diagram of the photocatalytic reactor using UV radiation is shown in Fig. 7.14. An aqueous solution of organic dye (100–200 ppm) is taken in the reactor and 100 mg of TiO_2 is added to it. The mixture is stirred, and the lamp switched on to initiate the photcatalytic oxidation reaction of the dye. The decrease in the concentration of the dye is measured by UV–Visible spectroscopy. A number of dyes like methylene blue (MB), orange G, alizarin S, methyl red, and congo red and organic compounds like *p*-nitrophenol, phenol, and salicylic acid have been tested for their photocatalytic activity with the combustion synthesized nano-TiO_2.[7–10]

Degradation of methylene blue has also been carried out under solar irradiation over different samples of titania prepared by solution combustion.

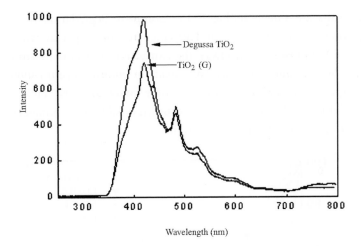

Fig. 7.13.   The photoluminescence (PL) emission spectra of combustion synthesized and commercial TiO$_2$ nanoparticles.

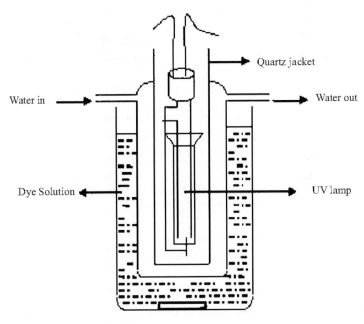

Fig. 7.14.   A schematic diagram of photo-catalytic reactor used for investigating the photocatalytic property.

These results are shown in Fig. 7.15. Complete degradation of 100 ppm dye is observed in 220 min with TiO_2 prepared from glycine while in the case of TiO_2 prepared from HMT the conversion is essentially complete at 300 min. While nanocrystalline TiO_2 made by the combustion shows complete degradation, the degradation reaction for commercial TiO_2 stops at 60 ppm. This may be because MB adsorbs irreversibly on commercial titania leading to saturation during degradation.

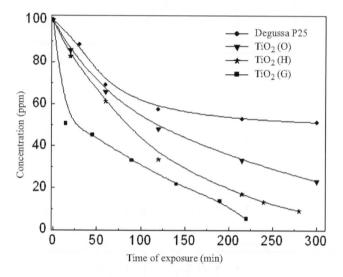

Fig. 7.15.   Degradation profile of methylene blue (MB) carried out under solar irradiation over different titania prepared by solution combustion method.

The reasons for higher photocatalytic activity of combustion-synthesized titania when compared to commercial TiO_2 are

(a) Crystalline nature of the nanoparticles.
(b) Hydrophilic nature of the surface (absorbing more water molecules as seen from the TGA).
(c) Nonadsorption of the dye molecules on the surface.
(d) Absorption in the visible region at about 500 nm.
(e) Lower band gap (2.5 eV).

It is gratifying to note that the same combustion-synthesized TiO_2 is reusable for photocatalytic activity. Additionally, combustion-synthesized titania degrades organic compounds at a faster rate compared to the degradation rates with commercial catalysts.[9]

7.4 METAL-ION-SUBSTITUTED TiO$_2$

A number of strategies have been proposed to improve the light absorption capability of TiO$_2$ and increase its carrier lifetime. It has been hypothesized that the addition of transition metals (M) to titania increases the rate of photocatalytic oxidation, because of electron scavenging by the metal ions at the surface of the semiconductor. This is achieved by the following reaction:

$$M^{n+} + e_{CB}^{-} \longrightarrow M^{(n-1)+} \tag{13}$$

This reaction prevents electron–hole recombination and increases the rate of formation of OH$^{\bullet}$ radical. Alternatively, nano-metal crystallites such as Ag, Ni, Pd impregnated over TiO$_2$ show higher photocatalytic activity mainly due to the electrons excited to the conduction band gathered by the metal particles, which in turn transfer them to the molecules.[11,12] Transition metal ions are doped for Ti ion in TiO$_2$ without difficulty by solution combustion method, leaving oxide ion vacancies for charge compensation. Metal-ion-substituted titania (Ag, Ce, Cu, Fe, V, W, and Zr) synthesized by solution combustion method have been investigated and compared with metal-impregnated titania for their photocatalytic properties.

7.4.1 *Synthesis and Photocatalytic Properties of Ti$_{1-x}$M$_x$O$_{2-\delta}$ (M = Ag, Ce, Cu, Fe, V, W, and Zr)*

Metal-substituted TiO$_2$ is prepared by solution combustion method taking precursors such as AgNO$_3$ for Ag, (NH$_4$)$_2$Ce(NO$_3$)$_6$ for Ce, NH$_4$VO$_3$ for V, Cu(NO$_3$)$_2$ for Cu , Fe(NO$_3$)$_3$·9H$_2$O for Fe, H$_2$WO$_4$ for W and ZrO(NO$_3$)$_2$·H$_2$O for Zr.[13] The dopant concentrations vary between 1 and 7.5 atom%. It is important to retain the parent structure of TiO$_2$ on substitution of transition metal ion in the lattice. The combustion mixture for the preparation of 5 atom% Fe/TiO$_2$ contains titanyl nitrate, ferric nitrate, and glycine in the mole ratio of 0.95:0.05:1.14.

Metal-ion-substituted titania shows characteristic color of the metal ion in the oxide matrix. The color of metal-substituted titania along with their particulate properties are summarized in Table 7.4.

Table 7.4.   Particulate properties of metal-ion-substituted TiO_2.

$Ti_{1-x}M_xO_{2-\delta}$	Color	Surface area $(m^2\,g^{-1})$	Crystallite size (nm)
TiO_2	Yellow	156	6
1 atom% W/TiO_2	Pale yellow	141	9
2 atom% W/TiO_2		119	10
2.5 atom% V/TiO_2	Brown	137	10
5 atom% V/TiO_2		110	11
2.5 atom% Ce/TiO_2	Dark yellow	97	6
5 atom% Ce/TiO_2		130	5
7.5 atom% Ce/TiO_2		127	5
2.5 atom% Zr/TiO_2	Dark yellow	113	8
5 atom% Zr/TiO_2		133	7
7.5 atom% Zr/TiO_2		146	8
2.5 atom% Fe/TiO_2	Brown	68	9
5 atom% Fe/TiO_2		64	10
7.5 atom% Fe/TiO_2		58	9
7.5 atom% Cu/TiO_2	Green	133	7
1 atom% Ag/TiO_2	Gray	135	7
7.5 atom% Pd/TiO_2	Dark gray	128	8

The photochemical degradation of dyes and organic molecules using metal-ion-substituted titania is compared with those of metal-impregnated titania and pure titania. Only Ag-and Pd-ion-substituted TiO_2 shows higher activity when compared to undoped TiO_2; other metal-ion-substituted titania show lower photocatalytic activity. Decrease of photocatalytic activity of metal-ion-doped TiO_2 is due to addition of "d" levels of metal in the band gap of TiO_2 which act as electron traps. Increase in photocatalytic activity of Ag-ion-and Pd-ion-substituted TiO_2 results from the photoreduction of these ions creating metal atoms acting as scavengers of electrons.

Figure 7.16 shows the degradation of methylene blue (MB) using TiO_2, $Ti_{0.99}M_{0.1}O_{2-\delta}$, where M=Pt, Ag, Mn, Fe, Co, Ni, and Cu. Except for Ag-doped titania, all other metal-ion-substituted TiO_2 show a lower

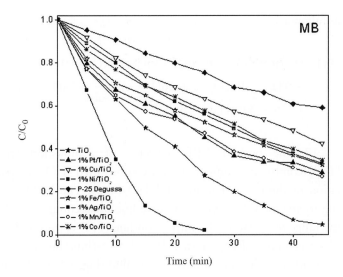

Fig. 7.16.   Degradation profile of MB using TiO_2, $Ti_{0.99}M_{0.1}O_{2-\delta}$ where M = Pt, Ag, Mn, Fe, Co, Ni, Cu.

photocatalytic activity compared to unsubstituted TiO_2. The initial degradation rate for Ag^+ is 30.6 μmol^{-1} s^{-1} whereas the unsubstituted combustion-synthesized TiO_2 gives an initial degradation rate of 21.6 μmol^{-1} s^{-1}. The initial degradation rates for Rhodamine B Base (RBB) with TiO_2 and 1% Ag/TiO_2 is 10.9 and 12.77 μmol^{-1}s^{-1}, respectively. The initial degradation rate of thionin acetate (TA) is 17.7 and 37.8 μmol^{-1} s^{-1} for TiO_2 and 1% Ag/ TiO_2, respectively. In other words, in 30 min 96% of the dye degrades with Ag^+-doped TiO_2 whereas only 56% dye degrades with undoped TiO_2.

7.4.2 Synthesis and Properties of $Ti_{1-x}Pd_xO_{2-\delta}$

Palladium-substituted titania ($Ti_{1-x}Pd_xO_{2-\delta}$) is prepared by taking stoichiometric amount of titanyl nitrate, palladium chloride and glycine. About 1 g of $TiO(NO_3)_2$ (in solution), 0.0095 g $PdCl_2$ (in solution), and 0.444 g glycine are taken in 300 ml borosilicate dish and heated in a muffle furnace maintained at 350°C.[14] The redox mixture ignites yielding a voluminous finely dispersed solid titania. Assuming complete combustion, the theoretical equation can be written as follows:

$$9(1 - x)TiO(NO_3)_2 + 10(1 - x)C_2H_5NO_2 + 9xPdCl_2$$
$$\longrightarrow 9Ti_{(1-x)}Pd_xO_{2-x} + 20(1 - x)CO_2 + 14(1 - x)N_2$$
$$+ (25 - 34x)H_2O + 18HCl \qquad (14)$$

In Fig. 7.17, the powder XRD patterns of combustion-synthesized 1% $Pd/TiO_2(comb)$ are compared to Pd-substituted TiO_2 prepared by impregnation method, i.e., 1 atom% $Pd/TiO_2(imp)$. The XRD peaks corresponding to anatase phase show extensive line broadening with crystallite sizes in the range of 8–9 nm. The diffraction line (111) corresponding to Pd metal is observed in the XRD pattern of 1% $Pd/TiO_2(imp)$ but not seen in the compound made by solution combustion method (Fig. 7.17). This indicates the substitution of Pd in TiO_2 in case of combustion-synthesized Pd/TiO_2 the formula of which is $Ti_{1-x}Pd_xO_{2-\delta}$.

The photoelectron spectra of $Ti_{1-x}Pd_xO_{2-\delta}$ show binding energies of $Pd(3d_{5/2,3/2})$ at 337.2 and 342.6 eV demonstrating Pd in 1 atom% $Pd/TiO_2(comb)$ in 2+ ionic state. Conversely, in 1 atom% $Pd/TiO_2(imp)$, binding energies of $Pd(3d_{5/2,3/2})$ are at 335 and 340.5 eV, respectively, confirming zero valent Pd in this compound. As Pd is in 2+ state in the Ti^{4+} site, there has to be an oxide ion vacancy for charge balance. For each Pd^{2+} ion, there is at least one oxide ion vacancy. So, the exact formula of 1 atom% $Pd/TiO_2(comb)$ is $Ti_{0.99}Pd_{0.01}O_{1.99}$. The Ti (2p) core level spectra of unsubstituted TiO_2 and $Ti_{0.99}Pd_{0.01}O_{1.99}$ show a Ti ($2p_{3/2}$) peak at 459 eV. Similar line shape and same line width shows Ti in 4+ state in $Ti_{0.99}Pd_{0.01}O_{1.99}$.

7.4.3 Catalytic Properties of $Ti_{1-x}Pd_xO_{2-\delta}$

The catalytic property of $Ti_{1-x}Pd_xO_{2-\delta}$ is investigated by TPR studies. The CO oxidation by O_2 over $Ti_{0.99}Pd_{0.01}O_{1.99}$, 1% $Pd/TiO_2(imp)$[14] and unsubstituted TiO_2 (150 mg each) is carried out with an inlet gas flow of $CO : O_2 = 2 : 2$ and a total gas flow of He at 100 cc min^{-1}. Figure 7.18 shows comparison of the catalysts in a light-off curve. A 100% CO conversion takes place below 100°C over $Ti_{0.99}Pd_{0.01}O_{1.99}$; T_{50} of 75°C is lower compared to 1% $Pd/TiO_2(imp)$ at 140°C. CO oxidation reaction rate is found

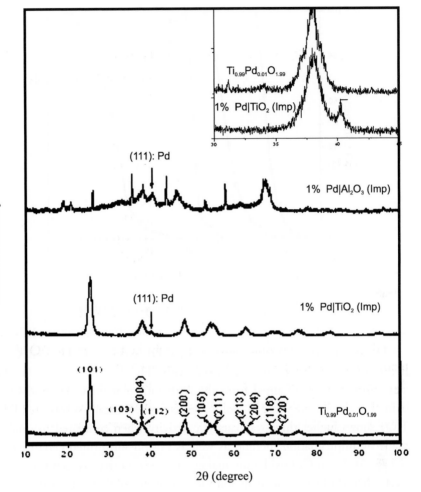

Fig. 7.17.   Powder XRD patterns of 1% Pd/TiO$_2$ and 1 atom% Pd/TiO$_2$(imp).

to be particularly high at low temperatures when compared to other ceria-supported catalysts presented in Chap. 5. Pd^{2+} ion is a potential adsorbent of CO in Ti$_{0.99}$Pd$_{0.01}$O$_{1.99}$.

The conversion of NO by CO over Ti$_{0.99}$Pd$_{0.01}$O$_{1.99}$ with equimolar mixture of 1:1 vol% NO and CO is carried out with excess oxygen. TPR profile of NO + CO + O$_2$ reaction and the percentage selectivity of N$_2$ and N$_2$O with temperature are presented in Fig. 7.19. The TPR profile clearly shows higher N$_2$ selectivity of this catalyst.

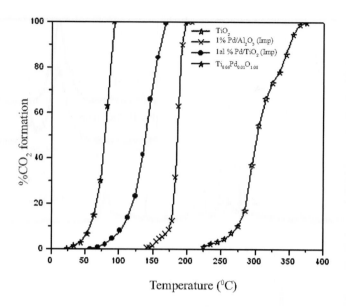

Fig. 7.18.   Comparison of the catalysts in a light-off curve.

To evaluate the TWC activity of $Ti_{0.99}Pd_{0.01}O_{1.99}$, C_2H_2 and C_2H_4 hydrocarbon oxidation are carried out with "HC":O_2 in 0.5:3 vol.%. The activation energies obtained from the Arrhenius plot for C_2H_2 and C_2H_4 oxidation are 70 and 53.42 kJ mol^{-1}, respectively. Ionically substituted Pd in TiO_2 shows higher three-way catalytic activity when compared to catalysts made by impregnation method as well as Pd-ion-substituted ceria catalysts (Chap. 5). From CO oxidation analogy it is proven that oxygen is dissociatively chemisorbed in the oxide ion vacancy of the catalyst. The hydrocarbon oxidation is also expected to be facilitated by the same phenomenon and indeed, lower activation energy is obtained with this catalyst. The oxide ion vacancy is found to be crucial for very high rates of CO and hydrocarbon oxidation with lowering activation energy. For NO + CO reaction, high N_2 selectivity with sufficiently high rate is observed. Dissociation of O_2 or NO (oxidant) in the oxide ion vacancy and adsorption of CO or hydrocarbon molecules (reductants) leading to bifunctional catalysis is the key for higher performance of these catalysts. The higher catalytic activity of $Ti_{1-x}Pd_xO_{2-\delta}$ compared to $Ce_{1-x}Pd_xO_{2-\delta}$ is due to the easy reducibility of Ti^{4+} ion compared to Ce^{4+} ion.

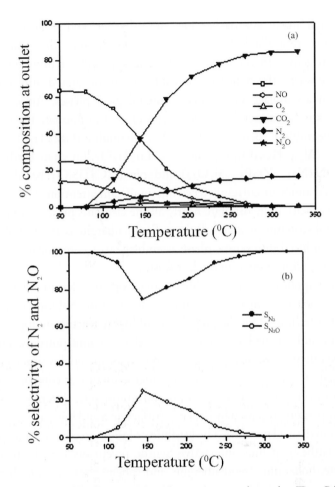

Fig. 7.19.   (a) TPR profile of NO + CO + O$_2$ reaction over the catalyst Ti$_{0.99}$Pd$_{0.01}$O$_{1.99}$ and (b) Percentage selectivity of N$_2$ and N$_2$O with temperature.

7.5   TITANATES FOR NUCLEAR WASTE IMMOBILIZATION

A major challenge confronting the nuclear industry is to provide an enduring solution to the problem of high-level radioactive liquid wastes (HLW) disposal. HLW is generated during the reprocessing of spent nuclear fuel rods which have radioactive half-life ($t_{1/2}$) of $10–10^6$ years and consequently need to be isolated from entering into the biosphere.

Although many approaches like multi-barrier, and glass and ceramic waste forms are available for the nuclear waste disposal, the ceramic waste form is attractive. In the ceramic synthetic approach of immobilizing the radioactive waste, the radioactive nuclides are incorporated into solid solution in an assemblage of synthetic mineralogical phases. Among various ceramic waste forms, synthetic rock (synroc) — a multiphase titanate-based ceramic assemblage (perovskite $CaTiO_3$), zirconolite ($CaZrTi_2O_7$), hollandite ($Ba_{1.23}Al_{2.46}Ti_{5.54}O_{16}$), pyrochlore ($Ln_2Ti_2O_7$, Ln = rare earth metals), NASICON ($Na_{1-x}Zr_2P_{3-x}Si_xO_{12}$), and sphene ($CaTiSiO_5$) based titanosilicate glass ceramics are currently receiving significant attention. These minerals have the capacity to incorporate nearly all the elements present in HLW into their crystal structure as solid solutions. Freudenbergite is an ideal phase that can be introduced into synroc phase assemblage to accommodate substantial amount of sodium present in nuclear waste.[15]

Nuclear applications need highly reactive oxide powders in order to assimilate the radioactive nuclei readily into their structure. Solution combustion method has been used for the preparation of highly reactive multiphase waste free synroc and sphene-based titanosilicate glass–ceramic powders. The combustion synthesized waste free Synroc-B precursor material is sinteractive to a high density (>95%) with fine-grained microstructure at comparatively low temperatures (1250°C) by cold pressing and atmospheric sintering. Lower sintering temperature is caused by the large surface area and fine particle nature of combustion derived powders. The feasibility of solution combustion approach for the synthesis of fine sinteractive Synroc precursors namely, perovskite, zirconolite, hollandite, freudenbergite, and sphene glass–ceramics have been clearly demonstrated. They are potential forms for nuclear waste immobilization and can help protect our environment from HLW. Combustion-derived nanopowders are not only sinteractive at low temperature but also show good chemical and mechanical compatibility and hydrothermal stability.

Synthesis of calcium titanate ($CaTiO_3$) is described as a representative combustion derived oxide material used for nuclear waste immobilization. A redox mixture of 5 g calcium nitrate, 4 g titanyl nitrate, and 3 g TFTA are dissolved in minimum amount of water in a Pyrex dish. The dish is introduced into a muffle furnace maintained at 450°C. The solution undergoes dehydration and ignites to burn yielding voluminous calcium titanate powder in less than 5 min. Similarly, calcium zirconolite, hollandite, SYNROC, and sphene (using three different fuels) are also prepared using combustion process.[15–18] Actual compositions of the redox mixture used for the combustion synthesis of SYNROC materials are summarized in Table 7.5.

Theoretical equations for the formation of calcium titanate, calcium zirconolite, hollandite, and calcium titanosilicates (sphene) by the combustion reactions may be written as

$$7Ca(NO_3)_2(aq) + 7TiO(NO_3)_2(aq) + 5\underset{TFTA}{C_4H_{16}N_6O_2}(aq)$$

$$\longrightarrow 7CaTiO_3(s) + 20CO_2(g) + 40H_2O(g) + 29N_2(g)$$

$$(12.7 \text{ mol of gases/mol of } CaTiO_3) \qquad (15)$$

$$4Ca(NO_3)_2(aq) + 4Zr(NO_3)_4(aq) + 8TiO(NO_3)_2(aq) + 25\underset{CH}{CH_6N_4O}(aq)$$

$$\longrightarrow 4CaZrTi_2O_7(s) + 25CO_2(g) + 75H_2O(g) + 70N_2$$

$$(42.5 \text{ mol of gases/mol of } CaZrTi_2O_7) \qquad (16)$$

$$2Ba(NO_3)_2(aq) + 4Al(NO_3)_3(aq) + 12TiO(NO_3)_2(aq) + 21\underset{CH}{CH_6N_4O}(aq)$$

$$\longrightarrow 2BaAl_2Ti_6O_{16}(s) + 21CO_2(g) + 63H_2O(g) + 59N_2(g)$$

$$(88 \text{ mol of gases/mol of } BaAl_2Ti_6O_{16}) \qquad (17)$$

$$2Ca(NO_3)_2(aq) + 2TiO(NO_3)_2(aq) + 2SiO_2(s) + 5\underset{CH}{CH_6N_4O}(aq)$$

$$\longrightarrow 2CaTiSiO_5(s) + 5CO_2(g) + 15H_2O(g) + 14N_2(g)$$

$$(17 \text{ mol of gases/mol of } CaTiSiO_5) \qquad (18)$$

Various intermediate phases formed during the calcinations are summarized in Table 7.4. The solid products of combustion are identified by their characteristic XRD patterns. Powder XRD patterns of combustion derived sphene (CH process) are shown in Fig. 7.20. The as-synthesized residue is X-ray amorphous and shows the presence of weakly crystalline CaO (Fig. 7.20a). At 600°C calcium titanate is the major phase along with CaO and TiO_2 formation (Fig. 7.20b). On further increase in the calcination temperature, $CaTiO_3$, CaO, and TiO_2 appear to react with SiO_2 to form $CaTiSiO_5$. Between 850°C

Table 7.5.   Compositions of redox mixtures and thermal phase evolution of combustion products.

Composition of redox mixtures	Compound	Calcination temp. (°C)	Phases
A (5 g) + B (4 g) + TFTA (2.7 g)	$CaTiO_3$	As-prepared	—
A (2 g) + B (3.48 g) + C (3.87 g) + CH (4.76 g)	$CaZrTi_2O_7$	As-prepared	TiO_2 (a) + ZrO_2 (t)
		700	TiO_2 (R) + ZrO_2 (R) + $CaTiO_3$ + $CaZrTi_2O_7$
		900	TiO_2 (R) + $CaZrTi_2O_7$
		1000	$CaZrTi_2O_7$
D (1 g) + B (0.24 g) + E (2.9 g) + CH (3.7 g)	Hollandite	As-prepared	Amorphous
		1000	Hollandite
A (1.84 g) + B (4.7 g) + C (1.02 g) + D (0.52 g) + E (2.48 g) + CH (2 g) + TFTA (2.1 g)	SYNROC	As-prepared	Amorphous + perovskite
		1100	Perovskite + zirconolite + hollandite
A (6.83 g) + B (5.43 g) + F (1.47 g) + CH (6.5 g)	$CaTiSiO_5$	As-prepared	Amorphous + CaO + TiO_2 (R)
		400	CaO + TiO_2 + $CaTiO_3$
		600	$CaTiO_3$ + CaO + TiO_2
		800	$CaTiSiO_5$ + $CaTiO_3$ + SiO_2
		875	$CaTiSiO_5$ + $CaTiO_3$ + SiO_2
		950	$CaTiSiO_5$ + $CaTiO_3$ + SiO_2
		1200	$CaTiSiO_5$
A (4.53 g) + B (3.6 g) + F (1.15 g) + TFTA (2.5 g)	$CaTiSiO_5$	1250	$CaTiSiO_5$
A (4.84 g) + B (3.85 g) + F (1.23 g) + DFH (9 g)	$CaTiSiO_5$	1200	$CaTiSiO_5$

A = $Ca(NO_3)_2$, B = $TiO(NO_3)_2$, C = $Zr(NO_3)_4$, D = $Ba(NO_3)_2$, E = $Al(NO_3)_3$, and F = Fumed SiO_2.

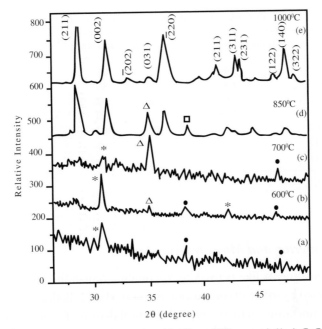

Fig. 7.20. XRD patterns of combustion derived sphene (CH process) (($*$) CaO, ($\bullet$) TiO_2, (Δ) $CaTiO_3$, and $\square$ α-crystoballite).

and 1000°C, α-crystoballite emerges out as a secondary intermediate phase (Fig. 7.20d). As the formation of $CaTiSiO_5$ progresses the amount of $CaTiO_3$ and SiO_2 decreases. The particulate properties of the combustion-derived oxide materials used for nuclear waste immobilization are summarized in Table 7.6.

Powder XRD patterns of as-synthesized zirconolite show the presence of anatase TiO_2 and tetragonal ZrO_2. At 700°C TiO_2, ZrO_2, and $CaTiO_3$ phases are observed along with a small amount of zirconolite. On further increase in the calcination temperature, $CaTiO_3$ appears to react with TiO_2 and ZrO_2 to form zirconolite. The formation of single-phase zirconolite occurs at 1000°C. The relatively low formation temperature (1000–1200°C) of zirconolite, hollandite, and sphene indicates the high reactivity of combustion products.

Table 7.6.   Particulate properties of combustion-derived oxide products.

Compound	Powder density (g cm^{-3})	Surface area (m^2 g^{-1})	Particle size[a] (nm)
CaTiO$_3$	3.46	21	80
CaZrTi$_2$O$_7$	2.8	39	50
Hollandite	3.94	37	40
SYNROC	2.98	9	130
Sphene/CH	2.57	67	30
Sphene/TFTA	2.83	65	30
Sphene/DFH	2.9	46	40

[a] From surface area.

7.5.1 *Sintering and Microstructure Studies*

Synroc constituents like perovskite, zirconolite, hollandite and sphene powders prepared by solution combustion method when pelletized and sintered at various temperatures achieve >90% theoretical density. The mechanical properties are summarized in Table 7.7. Typical microstructures of the sintered materials have been investigated (Fig. 7.21)

Table 7.7.   Sintering conditions and mechanical properties.

Conditions	Perovskite	Zirconolite	Hollandite	Sphene
Pressing load (MPa)	50	50	50	50
Sintering temperature (°C)	1200	1250	1250	1350
Sintering time (h)	1	2	2	2
Density * (g cm^{-3})	3.70 (92)	4.35 (98)	4.15 (96)	3.24 (92)
Apparent open porosity (%)	0.25	0.002	0.18	0.20
Vicker's hardness (GPa)	6.15	9.83	4.58	—
Thermal expansion coefficient $\alpha(x10^1 K^{-1})$ (RT to 1000K)	10.83	9.04	8.30	—

*Values in the parentheses are percentage theoretical density.

Fig. 7.21. SEM of (a) Perovskite (1200°C), (b) Zirconolite (1250°C), (c) Hollandite (1250°C), and (d) Sphene (1325°C).

7.6 CONCLUDING REMARKS

Combustion-synthesized nano-titania powders exhibit better photocatalytic property than commercial titania. Silver- and palladium-ion-substituted nano-titania show higher photocatalytic activity of oxidation of organic dyes like methylene blue, congo red, etc. Palladium-ion-substituted nano-titania is a highly efficient three-way catalyst. Synroc constituents like zirconolite, hollandite, and sphene prepared by solution combustion method show good sinteractivity indicating high reactivity for use in nuclear waste immobilization.

References

1. Fujishima A, Honda K, Electrochemical photolysis of water at a semiconductor electrode, *Nature* **238**: 37–38, 1972.
2. Frank SN, Bard AJ, Heterogeneous photocatalytic oxidation of cyanide and sulfite in aqueous solution at semiconductor powders, *J Phys Chem* **81**: 1484–1488, 1977.
3. Frank SN, Bard AJ, Heterogeneous photocatalytic oxidation of cyanide ion in aqueous solution at TiO_2 powder, *J Amer Chem Soc* **99**: 303–304, 1976.
4. Beck DD, Siegel RW, The dissociative adsorption of hydrogen sulfide over nanophase titanium dioxide, *J Mater Res* **7**: 2840–2845, 1992.
5. Aruna ST, Patil KC, Synthesis and properties of nanosize titania, *J Mater Synth Processing* **4**: 175–179, 1996.
6. Aruna ST, Combustion synthesis and properties of nanosize oxides: Studies on solid oxide fuel cell materials, PhD Thesis, Indian Institute of Science, Bangalore, India, 1998.
7. Nagaveni K, Hegde MS, Ravishankar N, Subbanna GN, Madras G, Synthesis and structure of nanocrystalline TiO_2 with lower band gap showing high photocatalytic activity, *Langmuir* **20**: 2900–2907, 2004.
8. Sivalingam G, Nagaveni K, Hegde MS, Madras G, Photocatalytic degradation of various dyes by combustion synthesized nano anatase TiO_2, *Appl Catal B: Environ* **45**: 23–38, 2003.
9. Nagaveni K, Sivalingam G, Hegde MS, Madras G, Photocatalytic degradation of organic compounds over combustion synthesized nano-TiO_2, *Environ Sci Technol* **38**: 1600–1604, 2004.
10. Nagaveni K, Sivalingam G, Hegde MS, Madras G, Solar photocatalytic degradation of dyes : high activity of combustion synthesized nano TiO_2, *Appl Catal B: Environ* **48**: 83–93, 2004.
11. Hirakawa T, Kamat PV, Charge separation and catalytic activity of Ag/TiO_2 core shell composite clusters under UV-irradiation, *J Amer Chem Soc* **127**: 3928–3934, 2005.
12. Hirakawa T, Kamat PV, Photoinduced electron storage and surface plasmon modulation in Ag/TiO_2 clusters. *Langmuir* **20**(14): 5645–5647, 2004.
13. Nagaveni K, Hegde MS, Madras G, Structure and photocatalytic activity of $Ti_{1-x}M_xO_{2\pm\delta}$ (M = W, V, Ce, Zr, Fe and Cu) synthesized by solution combustion method, *J Phys Chem. B* **108**: 20204–20212, 2004.
14. Roy S, Hegde MS, Ravishankar N, Giridhar M, Creation of redox adsorption sites by Pd^{2+} ion substitution in nano TiO_2 for high photocatalytic activity of CO oxidation, NO reduction, and NO decomposition, *J Phys Chem C* **111**: 8153–8160, 2007.

15. Muthuraman M, Arul Dhas N, Patil KC, Combustion synthesis of oxide materials for nuclear waste immobilization, *Bull Mater Sci* **17**: 977–987, 1994.

16. Muthuraman M, Patil KC, Senbagaraman S, Umarji AM, Sintering, microstructural and dilatometric studies of combustion synthesized synroc phases, *Mater Res Bull* **31**: 1375–1381, 1996.

17. Muthuraman M, Patil KC, Synthesis, properties, sintering and microstructure of sphene, CaTiSiO$_5$: A comparative study of coprecipitation, sol-gel and combustion processes, *Mater Res Bull* **33**: 655–661, 1998.

18. Muthuraman M, Patil KC, X-ray diffraction and ^{27}Al *MAS-NMR* spectroscopic studies on the evolution of synroc-B phases from the combustion residues (ash), *J Mater Sci Lett* **16**: 569–572, 1997.

Chapter 8

Zirconia and Related Oxide Materials

8.1 INTRODUCTION

Zirconia (ZrO_2) as an oxide material is hard (Vickers hardness 13–24 GPa), tough (fracture toughness 8–13 MPa m$^{1/2}$, strong (compressive strength 1000–1800 MPa), thermally stable (m.p. 2680°C), and chemically inert. It enhances the mechanical properties of other ceramics when uniformly dispersed in their matrices. Zirconia attracts attention due to its application as an engineering ceramic for both mechanical and electrical purposes.

To use zirconia to its full potential, certain cubic stabilizing oxides are added in varying amounts to significantly modify its properties. When added in sufficient amounts, these oxides form a fully stabilized zirconia or partially stabilized zirconia (PSZ) that exhibit a cubic structure from room temperature to its melting point.

Novel and innovative ceramic materials have been produced by additions of cubic stabilizing oxides particularly MgO, CaO, and Y_2O_3, resulting in considerable technological modification. The range of materials has been expanded by the use of specific rare earth additions, particularly cerium oxide which shows unusual "toughness". Zirconia-based engineering ceramics for instance, have now been developed to the stage where design of microstructure is possible by the control of composition, fabrication route, thermal treatment, and final machining.[1] Some applications of zirconia and related oxide materials are summarized in Table 8.1.

The synthesis and properties of nano-zirconia, stabilized zirconias, and composites of ZrO_2 with Al_2O_3, CeO_2, and TiO_2 are discussed in this

Table 8.1. Applications of zirconia and related oxide materials.

Materials	Applications
ZrO_2	Heat resistance lining in furnace, opacifier for ceramic glazes, glower in incandescent light, abrasives.
$ZrO_2/Cu–ZrO_2$	Catalyst for methanol synthesis
$CeO_2–ZrO_2$	Oxygen storage capacitor (OSC)
$ZrO_2–CaO–Ta_2O_5$	Electrode in MHD channel
Stabilized $ZrO_2–CaO$, MgO, Y_2O_3 (YSZ)	Solid electrolyte in fuel cells, sensors, water electrolyzers, O_2 semipermeable membrane, high-temperature heating elements, cathode in plasma torch
Ni–YSZ	Anode material for SOFC
$TiO_2–ZrO_2$	Dielectric resonator for microwave applications
$ZrSiO_4$ (zircon)	American diamond
Partially stabilized zirconia	Extrusion and wire drawing dies, hammers, knives, automotive parts, bearings, thermal barrier coatings
Rare earth doped ZrO_2/zircon	Glazes, gemstones, enamels and pigments for tiles, tableware, sanitaryware
$ZrO_2–Al_2O_3$ (ZTA)	Heat engines, rocket nozzles, cutting tools, grinding wheels
$Pb(Zr,Ti)O_3$ (PZT)	Gas igniters, recording pick ups, ultrasonic transducers, echo sounders, microphones and loud speakers
$Pb,La(Zr,Ti)O_3$ (PLZT)	Optical shutter
$Ln_2Zr_2O_7$, $CaTi_2ZrO_7$	Refractories, radio-active waste disposal, high-temperature heating elements, electrodes in MHD generators
$Ln_2(Eu)Zr_2O_7$	Host for fluorescent centres
NASICONs ($MZr_2P_3O_{12}$, M = Na, K, 1/2 Ca)	Solid electrolyte membranes for energy storage (batteries) energy conversion, electrochemical sensors, electrochromic displays, electrolysis of sodium chloride, immobilization of radio-active wastes, catalyst supports, spark plugs, heat exchangers, optical mirror substrates.

chapter. Synthesis and properties of NASICONS and pyrochlores have also been included.

8.2 ZIRCONIA

Zirconia exhibits three well-defined polymorphs: monoclinic (m), tetragonal (t), and cubic (c) phases. The monoclinic crystalline phase is stable at room temperature and transforms to the tetragonal phase with lattice contraction at 1170°C. On further heating at 2270°C, the tetragonal phase transforms to cubic phase with lattice expansion. On cooling, the reverse transformation occurs.

$$\text{monoclinic} \xleftrightarrow{\quad 950-1170°C \quad} \text{tetragonal}$$
$$\xleftrightarrow{\quad 1170-2270°C \quad} \text{cubic} \xleftrightarrow{\quad 2270-2680°C \quad} \text{melting point}$$

The structure of zirconia polymorph may be described by comparison with the cubic fluorite type (CaF_2) structure. The isometric projection based on the distorted fluorite structure of m-ZrO_2 is shown in Fig. 8.1(a). The monoclinic structure contains a number of interesting features. First, the zirconium (Zr) atom is in sevenfold coordination with the oxygen sublattice. The idealized Zr-coordinated polyhedron in baddeleyite is shown in Fig. 8.1(b). The O_{II} atoms in the cube are tetrahedrally coordinated ($Zr-O_{II} = 2.26\,\text{Å}$) while the O_I atoms on the other side of the cube are triangularly coordinated ($Zr-O_I = 2.04\,\text{Å}$). As a result, the deviation from the cubic fluorite structure is considerable, as illustrated by the distorted oxygen configuration. A second interesting feature is that the tetrahedrally bonded O_{II} layer has only one angle differing significantly from the tetrahedral angle (109.5°). This variation is due to the triangularly bonded O_I layer causing the

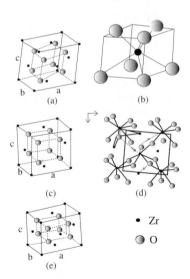

Fig. 8.1.   Schematic representation of the three zirconia polymorphs: (a) and (b) monoclinic, (c) and (d) tetragonal, and (e) cubic.

ZrO_7 groups to be tipped over. This group configuration results in a somewhat buckled nature of the O_{II} plane, while the O_I plane remains planar.

The distorted fluorite structure of tetragonal zirconia is shown in Fig. 8.1(c). The transformation of the monoclinic to tetragonal form is not straight forward. The most significant event occurring during this transformation is the change in coordination of Zr atoms from seven to eightfold. The probable atomic route by which this increase in coordination is achieved is indicated in Fig. 8.1(d). Above 2270°C, the t-ZrO_2 form transforms to the cubic zirconia having fluorite structure with *Fm3m* space group as shown in Fig. 8.1(e).

The unusual volume change, i.e., contraction on heating and expansion on cooling during t $\rightarrow$ m phase transformation leads to extensive cracking or even crumbling/shattering in sintered ZrO_2 material. If this aspect is not controlled or suppressed it would preclude the use of ZrO_2 from the field of advanced ceramics.

8.2.1 *Preparation and Properties of ZrO_2*

Monoclinic, tetragonal, and partially or fully stabilized zirconia powders are obtained by combustion of an aqueous solution containing zirconium oxy nitrate (ZON)/zirconium nitrate (ZN), and carbohydrazide in the required molar ratio of 1:1.125 and 1:2.5, respectively. In a typical experiment 5 g of zirconyl nitrate or 2.5 g of zirconium nitrate and 2.75 g of carbohydrazide dissolved in a minimum quantity of water are used for combustion. The pyrex container holding this solution when rapidly heated at 350 ± 10°C boils, foams, ignites to burn with a flame yielding voluminous foamy zirconia powders within 5 min (Fig. 8.2).[2]

The flame temperature measured by an optical pyrometer is 1400 ± 100°C for ZN–CH and 1100 ± 100°C for ZON–CH. Assuming complete combustion of zirconium oxynitrate/zirconium nitrate–CH redox mixture, the theoretical equation may be written as

$$4ZrO(NO_3)_2(aq) + 5CH_6N_4O \text{ (aq)}$$
$$\text{CH}$$

$$\xrightarrow{350°C} 4ZrO_2(s) + 5CO_2(g) + 15H_2O(g) + 14N_2(g)$$

$$(8.5 \text{ mol of gases/mol of } ZrO_2) \qquad (1)$$

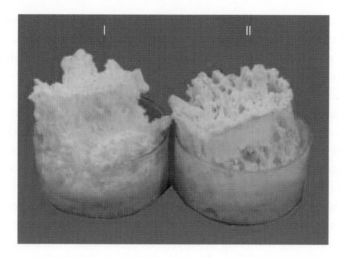

Fig. 8.2.   Foamy combustion product (I: mt-ZrO_2, II: m-ZrO_2).

$$2Zr(NO_3)_4(aq) + 5CH_6N_4O(aq)$$
$$CH$$
$$\xrightarrow{350°C} 2ZrO_2(s) + 5CO_2(g) + 15H_2O(g) + 14N_2(g)$$
$$(17 \text{ mol of gases/mol of } ZrO_2) \qquad (2)$$

The formation of zirconia is confirmed by powder XRD patterns which show a fully crystalline nature. ZN–CH redox mixture yields stable monoclinic (m) zirconia phase while ZON–CH shows characteristic XRD patterns of metastable tetragonal (mt) phase (Fig. 8.3).

ZrO_2 formed by combustion of zirconium oxynitrate is a low-temperature metastable tetragonal (mt-ZrO_2) phase probably due to the lower flame temperature (1100±100°C) of the redox reaction. The monoclinic phase gradually evolves with increasing calcination temperature above 1000°C. Metastable t-ZrO_2 on heating at 1150°C for 5 h converts to the stable monoclinic form. Formation of different phases of ZrO_2 by the combustion reaction may be explained by energetics and nucleation. The flame temperature of the zirconium nitrate and CH (ZN: CH, 1:2.5) redox mixture is very high (1400 ± 100°C) which essentially yields stable monoclinic zirconia. When lean fuel mixtures (ZN/CH ratio upto 1.25/1) are used for combustion, the exothermicity of the combustion reaction is lowered and metastable t-ZrO_2

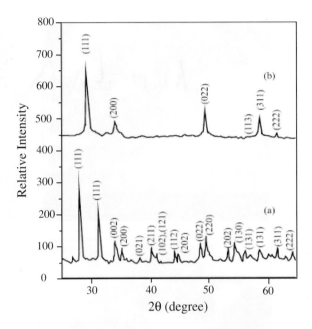

Fig. 8.3. XRD patterns of zirconia foam: (a) ZN–CH, (b) ZON–CH process.

phase appears aside from the m-ZrO$_2$. This indicates that the exothermicity (flame temperature) of the redox mixture plays an important role in the formation of the different phases of zirconia.

The various particulate properties of combustion-derived zirconia were determined by X-ray line broadening, surface area measurements, and transmission electron microscopy (TEM) studies (Fig. 8.4).

The typical properties of ZrO$_2$ synthesized from zirconium oxy nitrate/zirconium nitrate–CH are summarized in Table 8.2.

The powder densities of pure and stabilized ZrO$_2$ are in the range of 50–60% of the theoretical value and the surface areas are in the range of 4–13 m^2 g^{-1}. It is interesting to note that zirconia formed by combustion of zirconium nitrate and CH mixtures has a higher surface area than that formed by zirconium oxynitrate and CH mixture. This can be attributed to the greater amount of gases evolved per mole of ZrO$_2$ formed during combustion, i.e., 17 mol of gases evolved in case of zirconium nitrate when compared to 8.5 mol of gases generated in case of zirconium oxynitrate.

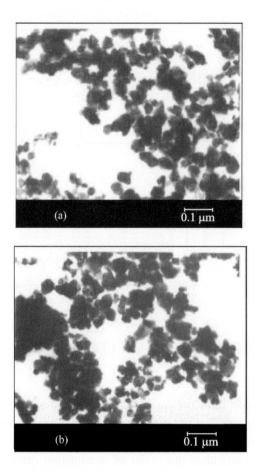

Fig. 8.4.   TEM of zirconia: (a) mt-ZrO_2 and (b) m-ZrO_2.

The combustion-derived zirconia powders are sinteractive. The microstructure of sintered ZON–CH at 1300°C for 1 h is shown in Fig. 8.5. For sintering studies, ZrO_2 powders obtained from zirconium oxy nitrate were selected as they yield mt-ZrO_2 phase. The SEM of combustion product shows the presence of coarse grains, intergranular porosity, and a less dense nature (84% of theoretical density). This is due to occurrence of mixed phases of mt + m-ZrO_2 and can be attributed to the volume change accompanied by the t–m phase transformation.

Nanosize zirconia is also prepared by the combustion of zirconium oxy nitrate and glycine in the presence of ammonium nitrate.[3] The theoretical

Table 8.2.   Properties of combustion-derived ZrO_2.

Properties	mt-ZrO_2 (ZON–CH)	m-ZrO_2 (ZN–CH)
Lattice constants	$a = b = 0.5821$ nm, $c = 0.51682$ nm	$a = 0.53134$ nm, $b = 0.52129$ nm, $c = 0.51516$ nm, $\beta = 99.17°$
Powder density (g cm^{-3})	3.2	3.0
Foam density (g cm^{-3})	0.02	0.01
Surface area (m^2 g^{-1})	4	13
Crystallite size[a] (nm)	30	35
Particle size[b] (μm)	0.46	0.15
Particle size[c] (μm)	0.2–0.5	0.2–0.5
Particle morphology	Platelet	Platelet

[a] From XRD.
[b] From surface area.
[c] From TEM.

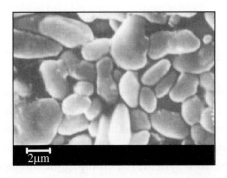

Fig. 8.5.   SEM of mt-ZrO_2 sintered at 1300°C for 1 h.

equation for the formation of ZrO_2 from reaction mixture used for the combustion is given below.

$$ZrO(NO_3)_2 + 2\underset{\text{Glycine}}{C_2H_5O_2N} + 4NH_4NO_3$$

$$\xrightarrow{350°C} ZrO_2(s) + 4CO_2(g) + 13H_2O(g) + 6N_2(g)$$

$$(23 \text{ mol of gases/mol of } ZrO_2) \qquad (3)$$

The TEM image of as-formed zirconia shows a uniform and compact distribution of the particles (Fig. 8.6). The particles show a nearly spherical to hexagonal geometry with ≤ 20 nm size. It appears that the particles are dispersed with negligible agglomeration and show surface area of 17 m^2 g^{-1}.

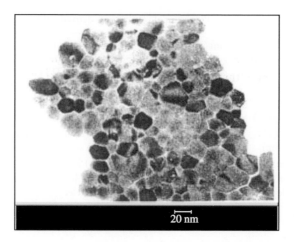

Fig. 8.6.   TEM of nanosize zirconia prepared by glycine fuel.

8.3 STABILIZED ZIRCONIA

As mentioned at the beginning of the chapter, sometimes the addition of a different ion to the unit cell of zirconia can induce dynamic changes in microstructure. The use of zirconia as an engineering ceramic is more feasible when the microstructure and the properties of interest are changed by the presence of another oxide system. For example, the monoclinic to tetragonal phase transformation of zirconia which occurs in the vicinity of the propagating cracks is martensitic in nature. This m $\rightarrow$ t transformation has led to the development of toughened zirconia where the role of the toughening agent is to retain the tetragonal phase at room temperature up to the operating temperature. The toughening process is dependent on the size, shape, and concentration of the t-phase. Retention of the t-ZrO_2 is accomplished by altering the chemistry either by adding other oxides or by controlling the particle size.[1]

Certain oxides such as yttria (Y_2O_3),[4] ceria (CeO_2), and less expensive MgO and CaO when added to zirconia dissolve to a significant extent because of their comparable atomic radii. They allow the stabilization of both the metastable tetragonal phase and cubic phase of zirconia. If insufficient stabilizing oxide is added then partially stabilized zirconia is produced rather than a fully stabilized form. In such a case the PSZ is made up of two or more intimately mixed phases, i.e., the cubic and tetragonal. It is possible to transform the tetragonal to monoclinic on cooling if stabilization is not complete.

Tetragonal stabilized zirconia polycrystal (TZP) and PSZ ceramics have received special attention because of their excellent mechanical properties like high fracture strength ($\sim$1500 MPa) and toughness ($\sim$13 MPa m$^{1/2}$). Stabilized zirconia ceramics are found to be useful in high-temperature solid oxide fuel cells (SOFC), high-temperature pH sensors, oxygen sensors, water electrolyzers, refractory materials, thermal barrier coatings, and as abrasives.

Partially and fully stabilized zirconias are prepared by the combustion of aqueous solutions containing zirconium nitrate, corresponding metal nitrate salts like Y_2O_3, MgO, CaO, CeO_2, etc., as stabilizers, and carbohydrazide fuel in the required mole ratio. The added stabilizer (metal nitrate) decomposes into the respective metal oxide (MO) and reacts with the metathetically formed ZrO_2 to give a solid solution. In this manner the stabilization of the high-temperature zirconia phase is achieved.

$$(1 - x)ZrO_2 + xMO \longrightarrow Zr_{1-x}M_xO_{2-x} \qquad (4)$$

The property of the stabilized zirconia ($Zr_{1-x}M_xO_{2-x}$) varies depending upon the type and concentration of the additive.

8.3.1 *Magnesia-Stabilized Zirconia*

Magnesia-stabilized zirconia powders (7 mol%, 9 mol%, and 14 mol% MgO) are prepared by the combustion of an aqueous solution containing stoichiometric amounts of zirconyl nitrate, magnesium nitrate, and CH in a cylindrical Pyrex dish.[5] The solution when heated rapidly at 350°C boils, foams and ignites to burn with a flame yielding voluminous foamy zirconia powder in less than 5 min. During combustion, the magnesium nitrate added to the redox mixture decomposes to MgO and stabilizes the high-temperature ZrO_2 phase.

The equivalence ratio for zirconium oxynitrate/magnesium nitrate and CH is 1:1.25. The combustion product shows the fully crystalline nature of zirconia powders (Fig. 8.7). ZrO_2–7 mol% MgO partitions into equilibrium phases at higher temperatures, following the sequence:

$$\text{t-}ZrO_2 \longrightarrow \text{t-}ZrO_2 + \text{m-}ZrO_2 + MgO \longrightarrow \text{m-}ZrO_2 + MgO \qquad (5)$$

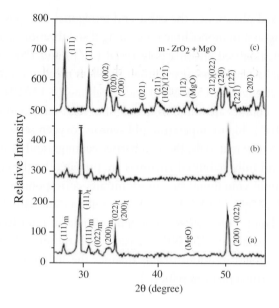

Fig. 8.7.   XRD patterns of calcined ZrO_2 –7 mol% MgO: (a) 950°C, (b) 1150°C, and (c) 1500°C.

As the calcination temperature is increased, the concentration of monoclinic phase also increases. The t-phase disappears only after the specimen is heated to 1500°C.

The variation in the lattice parameters of ZrO_2–7 mol% MgO with calcination temperature are calculated using the Garvie and Nicholson formula (Table 8.3). The data reveal that the cell parameter values decrease as the calcination temperature of the combustion residue increases.

Table 8.3.   Variations of cell parameters with temperature.

Temperature (°C)	Lattice parameters		c/a	Monoclinic phase (%)
	a (nm)	c (nm)		
As-formed	0.51331	0.51558	1.006	—
950	0.50974	0.51420	1.009	14
1150	0.50953	0.51350	1.007	15
1250	0.50935	0.51280	1.006	17

With an increase in the calcination temperature, the tetragonality (c/a) initially increases and then it decreases as the percentage of MgO content increases. The properties of ZrO_2–MgO powders are summarized in Table 8.4.

The powder densities of ZrO_2 are in the range of 50–60% of theoretical density and the surface areas in the range of 8–$14\,m^2\,g^{-1}$. As the MgO content increases there is a slight increase in the surface area while the crystallite size decrease. This occurs possibly due to the increase in the defect concentration indicating single-phase ($Zr_{1-x}Mg_xO_{2-x}$) formation of the combustion residue. In MgO-stabilized zirconia, the phase separation (MgO and m-ZrO_2) at high temperature reduces the densification, but when combustion-derived ZrO_2–MgO powders are cold pressed and sintered at $1300°C$ for $3\,h$ and rapidly cooled to room temperature, 98% of the theoretical density is achieved with a fine-grained microstructure (Fig. 8.8).

Table 8.4.   Particulate properties of ZrO_2–MgO powders.

Material	Phases by XRD	Crystallite size (nm)	Powder density ($g\,cm^{-3}$)	Surface area ($m^2\,g^{-1}$)	Particle size[a] (nm)
7 mol% MgO stabilized ZrO_2	t	29.0	3.2	8	234
9 mol% MgO stabilized ZrO_2	t + c	26.0	3.3	10	181
14 mol% MgO stabilized ZrO_2	c	24.0	3.4	14	126

t — tetragonal, c — cubic
[a] From surface area.

8.3.2 *Calcia-Stabilized Zirconia*

Calcia (10–15 mol%)-stabilized zirconia are prepared by the combustion of the aqueous solutions of their nitrates with zirconyl nitrate and carbohydrazide redox mixture. For example, a saturated aqueous solution of the redox mixture containing zirconyl nitrate (0.04 mol), calcium nitrate 0.0045 mol/(10 mol%)/0.0054 mol (12 mol%)/0.007 mol (15 mol%), and 0.04 mol carbohydrazide is introduced into a furnace maintained at $350°C$. Initially, the solution boils, foams, and then ignites to burn with flame yielding voluminous and foamy calcia-stabilized zirconia (CSZ) powder.

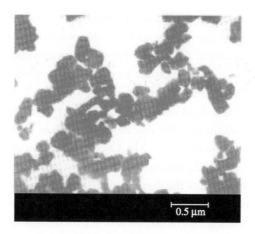

Fig. 8.8.   TEM of ZrO_2 –7 mol% MgO.

The XRD patterns of various CSZ samples indicate that 9 and 10 mol% calcia-doped zirconia samples are a mixture of monoclinic (m) and cubic (c) phases while 12–15 mol% calcia-doped zirconia samples exhibit cubic–fluorite phases. The scanning electron micrographs of CSZ are shown in Fig. 8.9. It is seen that in case of CSZ samples, the grain-size increases with dopant concentration. The oxide-ion conductivity of various CSZ samples in the temperature range of 347–800°C have been studied.[6]

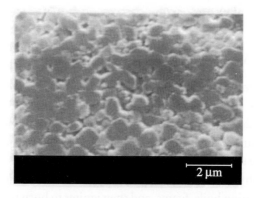

Fig. 8.9.   SEM of sintered calcia-stabilized zirconia.

8.3.3 *Yttria-Stabilized Zirconia (YSZ)*

Various phases of yttria (Y_2O_3)-stabilized zirconias with different mol% of yttria have been prepared as shown in Table 8.5. YSZ (3–10 mol%) is prepared by using zirconyl nitrate, yttrium nitrate, and carbohydrazide as fuel.[6] According to the phase diagram, the tetragonal zirconia phase (TZP) is stabilized by the addition of 2–3 mol% of yttria. The tetragonal cubic phase exists between 3.5 and 6 mol% of yttria and the cubic zirconia phase is stabilized by adding more than 6 mol% of yttria. The yttria-doped zirconia samples show phase stabilization of cubic fluorite at and above 8 mol% of yttria.

Table 8.5.   Particulate properties of YSZ.

Redox mixture	Phases by XRD	Powder density ($g\,cm^{-3}$)	Surface area ($m^2\,g^{-1}$)	Particle size[a] (nm)
ZON/ZN + 3 mol% Y_2O_3 + CH	t (t)	3.4 (3.8)	8.4 (5.4)	210 (292)
ZON/ZN + 4 mol% Y_2O_3 + CH	t + c (m + c)	3.4 (3.6)	3.5 (8.7)	504 (191)
ZON/ZN + 6 mol% Y_2O_3 + CH	c (c)	3.4 (3.6)	3.4 (9.4)	519 (177)

m — monoclinic, t — tetragonal, c — cubic. Values in parenthesis correspond to zirconia obtained by ZN and CH redox mixture.
[a] From surface area.

The various phases of YSZ powders formed are identified by powder XRD patterns (Fig. 8.10). Formation of zirconia solid solution with the additive is confirmed by the change in lattice parameters of ZrO_2 ($a = 0.50983$ nm, $c = 0.51758$ nm). The cell dimensions of stabilized zirconia are larger than pure zirconia (mt$-ZrO_2$: $a = 0.50821$ nm, $c = 0.51682$ nm). The lattice expansion implies that the additive goes into the zirconia lattice forming solid solution of the type ($Zr_{1-x}M_xO_{2-x}$).

The microstructures of sintered (1300°C for 1–4 h) CaO-, MgO-, CeO_2-, and Y_2O_3-stabilized ZrO_2 compacts have also been investigated.[2,7] It is observed that stabilized zirconia achieves 96–99% theoretical density with uniform fine grains (Fig. 8.11). The grain sizes and mechanical properties of TZP compacts are summarized in Table 8.6.

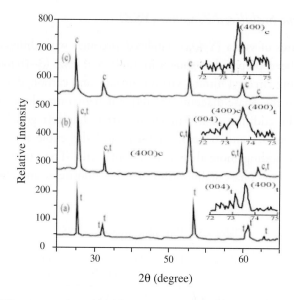

Fig. 8.10.   XRD patterns of yttria-stabilized zirconia (a) 3 mol% Y_2O_3, (b) 4 mol% Y_2O_3, and (c) 6 mol% Y_2O_3.

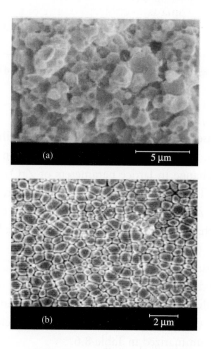

Fig. 8.11.   SEM of sintered (a) Ce-TZP and (b) Y-TZP.

Table 8.6.   Mechanical properties of TZP compacts.

Material	Sintering		Theoretical density (%)	Grain size (μm)	% 't' phase	Hardness (GPa)	Toughness (MPa m$^{-1/2}$)
	Temp. (°C)	Time (h)					
mt-ZrO$_2$	1300	1	84	1–5	50	—	—
12 mol% Ce-TZP	1300	3	98	1–3	100	13	10
3 mol% Y-TZP	1250	4	99	2–4	100	16	8
7 mol% Mg-TZP	1300	4	96	2–4	84	—	—
6 mol% Ca-TZP	1300	3	97	2–5	90	—	—

8.3.4   *Nickel in Yttria-Stabilized Zirconia (Ni–YSZ)*

The Ni cermet is generally used as the anode in solid oxide fuel cell (SOFC) with fully stabilized zirconia as electrolyte because of the low cost of Ni. The two-phase anode layer consists of YSZ component and metallic Ni that has a good electro-catalytic activity for H$_2$ oxidation. The nickel and YSZ systems should form continuous electronic and ionic pathways, respectively. The YSZ component in cermets is known to extend the effective electrolyte zone deep into the anode layer. This enables the oxide ions to be conveyed from the electrolyte to the three-phase boundary sites where the Ni electrode, YSZ component, and gas phase H$_2$ meet together.

In order to have good anodic activity for cermet electrodes, the cermet layer should have continuous network structure of both Ni and YSZ components, rich three-phase boundary sites and good adherence to the electrolyte. The conductivity of the cermet is strongly dependent on the microstructure and nickel content. The YSZ also provides an anode thermal expansion coefficient acceptably close to the other cell components. The requirement of anode material dictates the lowest nickel content possible, since Ni has the highest thermal expansion of the materials that comprise the SOFC.

Metal like nickel and cobalt are generally used in the form of cermets with YSZ because of the reducing conditions of the fuel gas prevailing during

SOFC operation. YSZ in cermets acts as a supporting matrix for metal particles, ensuring their uniform dispersion while preventing coalescing.

Nickel–YSZ cermet is usually made from YSZ and NiO powders. In SOFC, NiO is reduced *in situ* to nickel metal when exposed to the fuel. However, an area that has recently received serious attention is the long-term stability of the anode under the environments it is likely to face. It has been suggested that control of the preparation technique is a very important factor in controlling the electrode stability and performance. It is apparent that optimization of the anode involves increasing the Ni–YSZ gas boundary which can be achieved by decreasing the size of the Ni particles.[8] Nanosize Ni (15, 30, and 50 vol.%)– 9 mol% Y_2O_3-stabilized ZrO_2 (YSZ) cermets are prepared using the solution combustion method.

An aqueous solution containing stoichiometric amounts of yttrium nitrate, zirconyl nitrate, and nickel nitrate (oxidizers) and carbohydrazide (CH) fuel (O/F = 1) when introduced into a muffle furnace preheated to $350°C$, boils and forms a blue gel. A flame appears at one end and propagates throughout the mass in a very short duration (<1 min) yielding voluminous NiO–YSZ product. The reaction is violent and to avoid any spilling of the product from the container, a fuel lean (O/F = 2) redox mixture is employed for the synthesis to reduce the exothermicity of the reaction.

When fuel-rich (O/F = 1) conditions are used the as-formed powders are crystalline with corresponding NiO and cubic zirconia phases (Fig. 8.12). On the other hand, when a fuel-lean (O/F = 2) redox mixture is used the product that forms is weakly crystalline, which on calcination at $700°C$ shows peaks corresponding to cubic ZrO_2 and NiO. The extensive line-broadening observed in the XRD pattern confirms the nanosize nature of the powders. The crystallite sizes calculated from the XRD line broadening of $(111)_{ZrO_2}$ peak of the as-formed samples (O/F = 1) are in the range of 6–10 nm. Combustion-derived samples when subjected to reduction under flowing hydrogen at $800°C$ for 1 h convert NiO to Ni. After reduction, the crystallite sizes calculated from XRD line-broadening are found to be 12–17 nm.

The nanosize nature of the combustion-derived powders is further confirmed by TEM (Fig. 8.13). TEM images of the 30 Ni–YSZ reveals the presence of two phases (black and white grains) which are in the range of 5–12 nm. Their particulate properties are summarized in Table 8.7. With increase in the

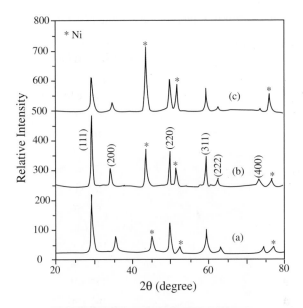

Fig. 8.12. XRD patterns of cermets sintered at 1350°C (a) 15 Ni–YSZ, (b) 30 Ni–YSZ, and (c) 50 Ni–YSZ.

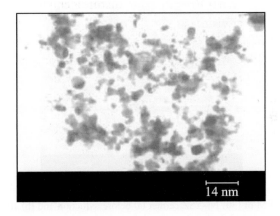

Fig. 8.13. TEM of 30 Ni–YSZ cermet reduced at 800°C.

vol.% of nickel in the cermet, the surface area decreases. This may be attributed to the increase in the flame temperature with increasing nickel content.

The SEM image of 30 Ni–YSZ (before and after reduction) are shown in Fig. 8.14. The dark areas in the micrographs correspond to the pores in the

Table 8.7.   Particulate properties of as-formed NiO–YSZ powders.

x vol.% Ni	Powder density (g cm^{-3})	Surface area (m^2 g^{-1})	Particle size (nm)
15	2.85	32	65
30	3.02	29	68
50	3.75	27	59

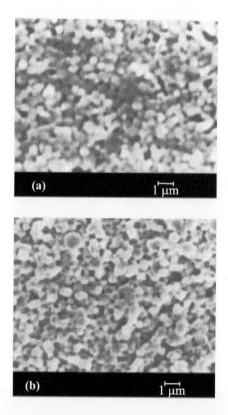

Fig. 8.14.   SEM of (30) Ni–YSZ cermet (a) before reduction and (b) after reduction.

sample, the gray areas are the nickel phase, and the white areas represents the zirconia phase.

Conductivity measurements were carried out for the Ni–YSZ cermet samples in a hydrogen atmosphere. The conductivity values for cermet samples with nickel contents of 15, 30, and 50 vol.% at 900°C were found to be

0.1, 40, and $989 \, S \, cm^{-1}$, respectively (Table 8.8). At a lower nickel content ($\leq 30 \, vol.\%$), conductivity increases with rise in temperature. Whereas with 50 vol.% the conductivity decreases with increasing temperature as Ni–YSZ cermet shows metallic behavior. The density of the sintered cermet pellets (1350°C) decreases after reduction due to the conversion of NiO into Ni.

Table 8.8.   Particulate properties of sintered Ni–YSZ (1350°C, 4 h).

x vol.% Ni	Density before reduction $(g \, cm^{-3})$	Density after reduction $(g \, cm^{-3})$	Conductivity $(S \, cm^{-1})$ 900°C	Thermal expansion coefficient at 900°C $(\times 10^{-6} \, K^{-1})$
15	6.08	5.50	0.103	10.40
30	5.97	4.84	40.000	11.64
50	5.33	4.58	989.000	13.20

The thermal expansion coefficient values (TEC) of nickel cermet increases linearly with the nickel content. The TEC of 30 NiO–YSZ is $11.64 \times 10^{-6} \, K^{-1}$ (Fig. 8.15), which compares favorably with that of the cathode and electrolyte materials ($10.70 \times 10^{-6} \, K^{-1}$).

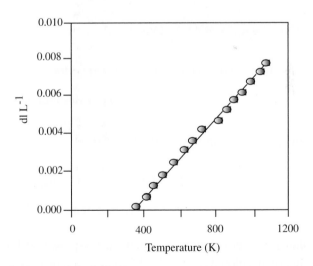

Fig. 8.15.   Thermal expansion versus temperature plot of 30 Ni–YSZ.

8.4 NANO-ZIRCONIA PIGMENTS

Zirconia-based pigments are important ceramic materials in the traditional ceramic industry. Zirconia pigments are used as glazes, gemstones, enamels, and coloring agent for tiles, tableware, and sanitary wares. Their use in colored ceramics is based on the ability to withstand the extreme chemical and thermal conditions encountered during firing processes used for glazing and enameling. Conventionally, preparation of colored zirconia is very difficult because of its high temperature of formation ($>1200°C$) which requires a mineralizer like sodium fluorite to give a homogeneous color. A number of zirconia-based pigments have been prepared without the addition of any mineralizer by the solution combustion method.[9]

Transition elements (V, Mn, Fe, Co) and rare earth elements (Ce, Pr, Nd) are used as dopants to make zirconia pigments. Various colored zirconia are prepared by mixing transition metal nitrates as coloring agents (0.3–1.0%) with ZON/ZN–CH redox mixture in aqueous medium and rapidly heating at 350°C. For the preparation of zircon, combustion was carried out in the presence of calculated amount of fumed silica. The products are voluminous, like ZrO_2.

Assuming complete combustion, theoretical equations for formation of metal-doped ZrO_2 and $ZrSiO_4$ can be written as follows:

$$4ZrO(NO_3)_2(aq) + 5CH_6N_4O(aq)$$
$$\text{CH}$$

$$\xrightarrow[350°C]{M^{x+}} 4M^{x+}/ZrO_2(s) + 5CO_2(g) + 15H_2O(g) + 14N_2(g)$$

$$(8.5 \text{ mol of gases/mol of } ZrO_2) \quad (6)$$

$$2Zr(NO_3)_4(aq) + 2SiO_2(s) + 5CH_6N_4O(aq)$$
$$\text{CH}$$

$$\xrightarrow[350°C]{M^{x+}} 2M^{x+}/ZrSiO_4(s) + 5CO_2(g) + 15H_2O(g) + 14N_2(g)$$

$$(17 \text{ mol of gases/mol of } ZrO_2) \quad (7)$$

The particulate properties of pigments obtained are summarized in Table 8.9.

The formation of colored zirconia by the combustion process demonstrates the high *in situ* flame temperature of the redox mixture. The dopants and

Table 8.9. Particulate properties of zirconia pigments.

Pigment	Powder density $(g\,cm^{-3})$	Surface area $(m^2\,g^{-1})$	Particle size (nm)
Zirconia (t-ZrO_2)	3.2	15	125
Zircon ($ZrSiO_4$)	4.9	66	19

the colors of combustion-derived zirconia and zircon powders are given in Table 8.10.

Table 8.10. Various colors of zirconia pigments.

Dopant	Metal (M^{x+})	Zirconia color	Zircon color[a]
NH_4VO_3	V	Pale yellow (0.3) Yellow (1.0)	Blue (1.5) Greenish blue (2.5)
$Mn(NO_3)_2 \cdot 4H_2O$	Mn	Light green (0.3) Pale violet (0.6)	Violet-gray (1.5)
$Fe(NO_3)_3 \cdot 9H_2O$	Fe	Light brown (0.3)	Pink (1.5)
$Co(NO_3)_2 \cdot 6H_2O$	Co	Pale violet (0.3) Brownish black (0.6)	Blue-violet (1.5)
$(NH_4)_2Ce(NO_3)_6 \cdot H_2O$	Ce	Pale orange (0.3)	
$Pr(NO_3)_3 \cdot 6H_2O$	Pr	Light brown (0.3)	Yellow (0.5)
$Nd(NO_3)_3 \cdot 6H_2O$	Nd	Pale violet (0.3)	

Values in parentheses correspond to the atom percentage of the dopant ion added in the redox mixture.
[a] After calcinations at 1300°C for 2 h except for V-doped zircon (900°C, 2 h)

The as-formed M^{x+}/ZrO_2 are colored. However, zircon pigment formation requires sintering of the products to $\sim$1300°C, probably due to the rigid nature of the zircon structure (Fig. 8.16). The color of these pigments has been attributed to the substitution of M^{x+} in ZrO_2 forming $M_xZr_{1-x}O_2$, whereas zircon pigments are formed by the inclusion of M^{x+} in $ZrSiO_4$ lattice.

Zirconia powders formed are X-ray crystalline corresponding to the tetragonal zirconia phase and due to the nanosize, extensive line-broadening is observed. The phase and color of zirconia were stable even after heating at a high temperature of 1300°C.[10]

The presence of dopant ion that imparts characteristic color in ZrO_2 is confirmed by UV–visible and fluorescence spectroscopy. For example, the

ZIRCON

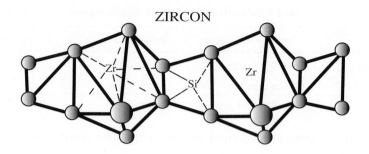

Fig. 8.16.   Structure of zircon.

electronic spectrum of as-prepared 0.3 atom% Mn–ZrO$_2$ shows a band at 382 nm corresponding to a $^6A_{1g} \rightarrow {}^4T_{1g}$ transition and a broad band having a maximum around 553 nm corresponding to a $^6A_{1g} \rightarrow {}^4T_{2g}$ transition of Mn^{2+}. The fluorescence spectrum of 0.3 atom% Mn–ZrO$_2$ (Fig. 8.17) shows the characteristic emission peak of Mn^{2+} at 470 nm, with less intense peaks at 450 and 480 nm. These factors explain the color of Mn–ZrO$_2$. When the

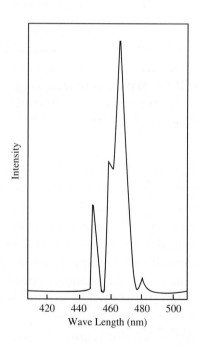

Fig. 8.17.   Fluorescence spectrum of 0.3 at.% Mn–ZrO$_2$.

concentration of Mn is low (0.3 atom% Mn) it stabilizes Mn^{2+} forming a solid solution with zirconia and giving a green color. As the Mn concentration is increased, the color changes to pale violet (at 0.6 atom% Mn) and this color change could be due to the stabilization of Mn^{4+} in the zirconia lattice. A further increase in Mn concentration (above 0.6 atom%) led to a black color due to the presence of mixed valences of manganese.

8.5 ZrO_2–Al_2O_3 SYSTEM: ZTA

Zirconia-toughened alumina (ZTA) is of interest because of its enhanced mechanical properties such as toughness, strength, and hardness. ZTA finds application in heat engines, rocket nozzles and cutting tools. The increased toughness in ZTA is attributed to the volume and shape changes that occur during the stress-induced metastable tetragonal to monoclinic transformation of zirconia. To yield optimum strength and toughness, certain design criteria must be fulfilled by zirconia particles so as to disperse in the alumina matrix. These are

- Retention of all zirconia particles in a metastable phase in the fine-grained α-alumina matrix before ceramic processing.
- Uniformity of the t-ZrO_2 particle distribution.
- A small and narrow size distribution of ZrO_2 particles.
- Absence of coarsening of t-ZrO_2 particles during heat treatments.

Zirconia–alumina (ZrO_2/Al_2O_3) composites containing different weight percentages of ZrO_2 (5, 10, 20, 30, 40, 50, and 80 wt.%) are prepared by the combustion of aqueous solutions containing stoichiometric amounts of aluminum nitrate, zirconyl nitrate, urea, or carbohydrazide.[11,12] Assuming complete combustion, theoretical equations for the reactions can be written as follows:

$$3ZrO(NO_3)_2(aq) + 5CH_4N_2O(aq)$$
$$\text{Urea}$$
$$\longrightarrow 3ZrO_2(s) + 5CO_2(g) + 10H_2O(g) + 8N_2(g)$$
$$(\sim 8 \text{ moles of gases/mole of } ZrO_2) \qquad (8)$$

$$2Al(NO_3)_3(aq) + 5CH_4N_2O(aq)$$
$$\underset{\text{Urea}}{}$$
$$\longrightarrow Al_2O_3(s) + 5CO_2(g) + 10H_2O(g) + 8N_2(g)$$
$$(23 \text{ moles of gases/mole of } Al_2O_3) \qquad (9)$$

In a typical experiment, 40 wt.% t-ZrO_2/Al_2O_3 composite is prepared by the combustion of an aqueous redox mixture containing 20 g aluminum nitrate, 2.53 g zirconyl nitrate, and 8.90 g urea. Combustion is carried out in a preheated muffle furnace (500±10°C) resulting in a flaming type reaction, comparable to the combustion reaction of α-Al_2O_3. Other compositions of t-ZrO_2/Al_2O_3 are similarly prepared. The actual weights for the preparation of various composites are given in Table 8.11.

The formation of t-ZrO_2/Al_2O_3 composites are confirmed by their characteristic XRD patterns. The as-derived zirconia–alumina composite powders are

Table 8.11.   Compositions of the redox mixtures and properties of ZTA powders.

Wt% of ZrO_2 in Al_2O_3	Composition of the redox mixtures		Powder density ($g\,cm^{-3}$)		Surface area ($m^2\,g^{-1}$)		Particle size[a] (nm)	
	U	CH	U	CH	U	CH	U	CH
5	20 g A + 0.3160 g Z + 8.12 g U	—	3.25	—	9.8	—	150	—
10	20 g A + 0.6328 g Z + 8.12 g U	7 g A + 0.24 g B + 2.78 g CH	3.32	2.8	8.3	63	210	34
20	20 g A + 1.2656 g Z + 8.45 g U	7 g A + 0.54 g B + 2.88 g CH	3.46	2.9	5.4	54	320	38
30	20 g A + 1.898 g Z + 8.67 g U	—	3.54	—	3.2	—	530	—
40	20 g A + 2.531 g Z + 8.90 g U	—	3.70	—	1.4	—	1100	—
50	—	7 g A + 2.16 g B + 3.45 g CH	—	3.2	—	47	—	39
80	—	7 g A + 8.64 g B + 5.7 g CH	—	3.7	—	40	—	40

A = $Al(NO_3)_3 \cdot 9H_2O$, Z = $ZrO(NO_3)_2 \cdot 3H_2O$, U = Urea, CH = Carbohydrazide.
[a] From surface area.

amorphous, which on calcinations at temperatures from $1000°C$ to $1400°C$ form the ZTA phase. The XRD patterns show characteristic peaks of α-Al_2O_3 along with those of t-ZrO_2. The evolution of various phases as a function of ZrO_2 content and temperature are shown in Figs. 8.18(a) and 8.18(b), respectivly. The absence of the two satellite peaks flanking the $30.3°$ peak of t-ZrO_2 clearly indicates the absence of the monoclinic phase. The relative intensity of t-ZrO_2 peak ($30.3°$) linearly increases with increase in the wt.% of ZrO_2 in these composites.

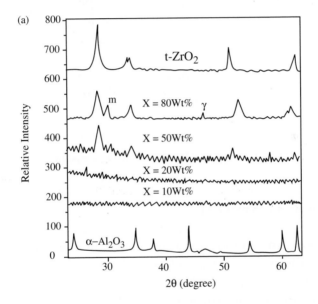

Fig. 8.18a.   XRD of $xZrO_2 - (1-x)Al_2O_3$ (m = monoclinic-ZrO_2; γ = γ-alumina).

Average crystallite size of ZrO_2 is 17.5 nm calculated from X-ray line broadening, which is well below the Gravie's critical size[13] for the transformation of tetragonal monoclinic phase of zirconia. Interestingly, it is observed that the metastable tetragonal phase of ZrO_2 is retained and the crystallite size does not increase even after several heating and cooling cycles at $1100°C$. The EPMA and TEM images of ZTA given in Figs. 8.19(A) and 8.19(B) show uniform distribution of intragranular ZrO_2 in the matrix of α-Al_2O_3.

Table 8.11 gives compositions for the combustion of the composites along with their particulate properties like density, surface area, and particle size.

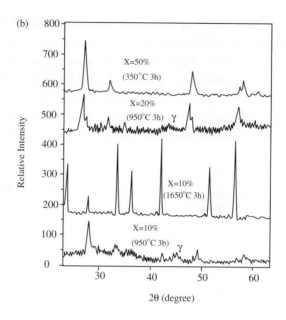

Fig. 8.18b.   XRD of calcined $xZrO_2 - (1 - x)Al_2O_3$ composites ($\gamma = \gamma$-alumina).

8.6 ZrO_2–CeO_2 SYSTEM

Ceria has been currently used as an active component in the three-way catalysts for automotive exhaust (oxidation of CO and hydrocarbons (HC) and reduction of NO_x) as discussed in Chap. 5. One of the important roles of CeO_2 in these multicomponent systems is to provide surface active sites; the other is to act as an oxygen storage/transport medium by shifting between Ce^{3+} and Ce^{4+} states under reducing and oxidizing conditions, respectively. These redox properties are strongly enhanced if foreign cations such as Zr, Gd, Pr, Tb, and Pb are introduced into the CeO_2 lattice by forming solid solutions.

The introduction of ZrO_2 into the CeO_2 lattice has recently been reported to strongly affect the reduction features of ceria. This occurs through structural modifications of the fluorite-type lattice of ceria as a consequence of the substitution of Ce^{4+} (ionic radius 0.97 Å) with Zr^{4+} (ionic radius 0.84 Å). The effect of substitution is to decrease the cell volume, thereby lowering the activation energy of oxygen-ion diffusion within the lattice and consequently favoring the reduction. The introduction of Zr also enhances the formation of

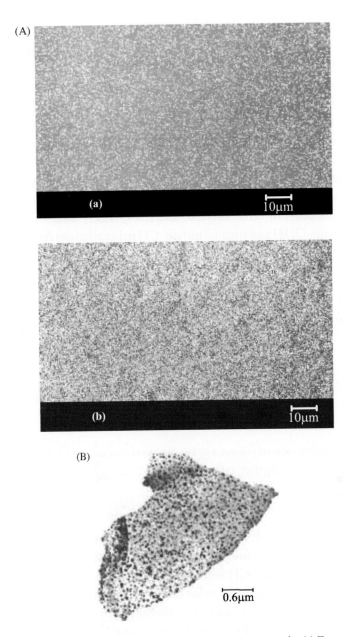

Fig. 8.19. (A) EPMA micrographs of a 10 wt.% ZTA composite powder (a) Zr mapping and (b) Al mapping. (B) TEM of as-prepared 40 wt.% ZTA composite powder.

structural defects which are expected to play an important role in determining the redox behavior. Thermal stability also increases with the addition of ZrO_2 into CeO_2 and helps to reduce surface area loss upon ageing.[14]

$Ce_{1-x}Zr_xO_2$ ($x = 0.2$–0.9) solid solution is synthesized by the solution combustion method using CH and ODH fuels. In a typical experiment, $Ce_{0.5}Zr_{0.5}O_2$ is prepared from an aqueous redox mixture containing cerous nitrate $Ce(NO_3)_3 \cdot 6H_2O$, zirconyl nitrate, and ODH + CH fuel in the molar ratio of 1:1: (1.875 + 1.25), which is introduced into a preheated muffle furnace ($350 \pm 10°C$). The solution boils and froths, followed by the appearance of flame yielding a voluminous product.

The formation of ceria, zirconia, and CeO_2–ZrO_2 solid solutions from cerous nitrate–ODH and zirconyl nitrate–CH redox mixtures can be written as

$$4Ce(NO_3)_3(aq) + 6C_2H_6N_4O_2 \text{ (aq)} + O_2(g)$$
$$\text{ODH}$$
$$\longrightarrow 4CeO_2(g) + 12CO_2(s) + 18H_2O(g) + 18N_2(g)$$
$$(12 \text{ mol of gases/mol of } CeO_2) \qquad (10)$$

$$4ZrO(NO_3)_2(aq) + 5CH_6N_4O(aq)$$
$$\text{CH}$$
$$\longrightarrow 4ZrO_2(s) + 5CO_2(g) + 15H_2O(g) + 14N_2(g)$$
$$(8.5 \text{ mol of gases/mol of } ZrO_2) \qquad (11)$$

Solution combustion of cerous nitrate, zirconyl nitrate, and CH/ODH redox mixtures in the appropriate ratio gives desired $Ce_{1-x}Zr_xO_2$ ($x = 0.2$–0.9).

$$(1 - x)CeO_2 + xZrO_2 \longrightarrow Ce_{1-x}Zr_xO_2 \qquad (12)$$

The combustion reaction is vigorous when CeO_2 content is more. The heats of combustion calculated from the heat of formation of reactants and products at STP for cerous nitrate–CH redox mixture is $-1906.23 \text{ kJ mol}^{-1}$ for CeO_2 and that of zirconyl nitrate–CH redox mixture is $-439.15 \text{ kJ mol}^{-1}$ for ZrO_2. Accordingly, in order to reduce the exothermicity of the cerous nitrate–CH combustion a fuel lean redox mixture containing 0.94 mol of CH is used.

The as-formed combustion products show broad XRD peaks, indicating nanocrystalline nature of the primary particles (Fig. 8.20). It is difficult to distinguish the different equilibrium phases present in the XRD pattern of

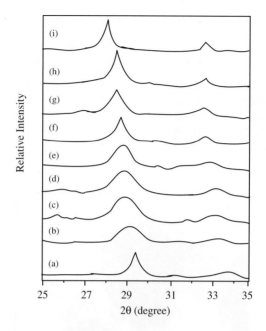

Fig. 8.20.   XRD patterns of as-formed $Ce_{1-x}Zr_xO_2$ (a) $x = 0.9$, (b) $x = 0.8$, (c) $x = 0.7$, (d) $x = 0.6$, (e) $x = 0.5$, (f) $x = 0.4$, (g) $x = 0.3$, (h) $x = 0.2$, and (i) $x = 0$.

the as-formed nanostructured combustion products because of line broadening. However, the XRD patterns of the calcined samples (1350°C/2 h) reveal three distinct regions of solid solutions: when $x < 0.3$ the XRD pattern corresponds to cubic $CeO_2(ZrO_2)$; when $x = 0.3–0.8$ it shows cubic $CeO_2(ZrO_2)$ and t-$ZrO_2(CeO_2)$ phases; above $x = 0.8$, a single t-ZrO_2 (CeO_2) phase is observed.

The as-formed $Ce_{0.1}Zr_{0.9}O_2$ is a metastable tetragonal solid solution, which transforms diffusionlessly into monoclinic phase during cooling. The particulate properties of the combustion derived ceria-zirconia solid solutions are summarized in Table 8.12.

The crystallite size of combustion-derived $CeO_2–ZrO_2$ solid solutions as calculated from XRD line broadening is in the range of 5–18 nm. This is further supported by TEM image of 10–12 nm (Fig. 8.21). These powders have a high surface area in the range of 36–120 $m^2 g^{-1}$, which decreases with increase in the concentration of ZrO_2.

Table 8.12.   Particulate properties of ceria–zirconia solid solutions.

Sample	Lattice constant a (nm)	Powder density $(g\,cm^{-3})$	Surface area $(m^2\,g^{-1})$	Particle size (nm)
$Ce_{0.8}Zr_{0.2}O_2$ (ODH)	0.541	2.4	120	18
$Ce_{0.6}Zr_{0.4}O_2$ (ODH)	0.530	2.7	90	24
$Ce_{0.5}Zr_{0.5}O_2$ (CH)	0.528	2.9	83	24
$Ce_{0.3}Zr_{0.7}O_2$ (CH)	0.524	3.1	71	27
$Ce_{0.2}Zr_{0.8}O_2$ (CH)	0.518	3.2	36	51

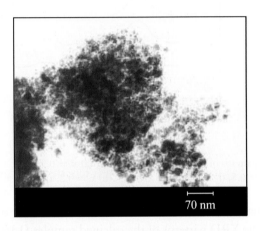

Fig. 8.21.   TEM of as-formed $Ce_{0.2}Zr_{0.8}O_2$.

8.7 ZrO_2–TiO_2 SYSTEM ($ZrTiO_4$ and $Zr_5Ti_7O_{24}$)

Zirconium titanate-based ceramics are widely used in electronic applications where low-loss, temperature-stable dielectric materials are required. Dielectric resonators are used for microwave components, microwave filters, and frequency-stable oscillators.

The redox mixtures containing a calculated amount of $TiO(NO_3)_2$ (TON), $Zr(NO_3)_4/ZrO(NO_3)_2$, and TFTA/CH are dissolved in minimum amount of water in a Pyrex dish and combustion is carried out in a muffle furnace maintained at $450 \pm 10°C$. The mixture boils, froths, and ignites to burn, producing a voluminous product in less than 5 min. The combustion is of smouldering type (flameless combustion).[15] Assuming complete combustion, the equation for the formation of zirconium titanate may be written as

$$4TiO(NO_3)_2(aq) + 4Zr(NO_3)_4(aq) + 15CH_6N_4O\,(aq)$$
$$\xrightarrow{\text{CH}} 4ZrTiO_4(s) + 15CO_2(g) + 45H_2O(g) + 42N_2(g)$$
$$(25.5\ \text{mol of gases/mol of } ZrTiO_4) \qquad (13)$$

$$2TiO(NO_3)_2(aq) + 2ZrO(NO_3)_2(aq) + 5CH_6N_4O\,(aq)$$
$$\xrightarrow{\text{CH}} 2ZrTiO_4(s) + 5CO_2(g) + 15H_2O(g) + 14N_2(g)$$
$$(17\ \text{mol of gases/mol of } ZrTiO_4) \qquad (14)$$

The composition of the redox mixture used for the preparation of ZrO_2:TiO_2 systems and the properties of the zirconium titanates are listed in Table 8.13.

The as-formed combustion products are weakly crystalline, and yield single-phase zirconium titanates on calcination at 1100–1200°C for 2 h. The XRD pattern of the as-formed product shows zirconium titanate along with a small amount of anatase titania. It has been reported that for the formation of $Zr_5Ti_7O_{24}$ phase Y_2O_3 is essential. However, the XRD pattern shows the

Table 8.13.   Compositions of redox mixtures and properties of zirconium titanates.

Composition of the redox mixtures	Powder density $(g\,cm^{-3})$	Surface area $(m^2\,g^{-1})$	Particle size (nm)
Zr:Ti = 1:1			
TON:ZN:TFTA = 1:1:1.08	3.49	53	30
TON:ZON:TFTA = 1:1:0.73	4.11	79	20
TON:ZN:CH = 1:1.376	3.96	13	120
TON:ZON:CH = 1:1:2.51	4.87	11	110
Zr:Ti = 5:7			
TON:ZN:TFTA = 1:0.71:0.87	4.11	46	30
TON:ZON:TFTA = 1:0.71:0.63	4.28	90	16
TON:ZN:CH = 1:0.71:3.04	4.60	15	80
TON:ZON:CH = 1:0.71:2.14	4.33	10	140

TON = $TiO(NO_3)_2$, ZN = $Zr(NO_3)_4$, ZON = $ZrO(NO_3)_2$.

presence of single phase $Zr_5Ti_7O_{24}$ at low calcination temperature even without the addition of Y_2O_3. These powders also have very high surface area in the range of $11–90 \, m^2 \, g^{-1}$.

The surface area of zirconium titanates prepared by CH fuel is very low when compared to TFTA fuel. The latter process involves smouldering combustion while the CH process is flaming type. This leads to partial sintering of the CH-derived products, reducing their surface area. To confirm this TON–ZN–CH process is carried out at various molar ratios of CH. The surface area of the product increases with decrease in the molar content of CH in the redox mixture. Also, a decrease in flame temperature is observed with decrease in the CH content. Zirconium titanate obtained from the combustion of fuel lean redox mixture with TON:ZN:CH in the molar ratio of 1:1:1.25 has a surface area of $22 \, m^2 \, g^{-1}$. The products derived from zirconium nitrate in the CH process have higher surface area than those derived from zirconyl nitrate.

Interestingly, in the case of TON–ZON–TFTA process, if the reaction mixture is stirred for 20 min it forms an opaque gel and the zirconium titanate obtained from the decomposition of the gel have very high surface area of up to $130 \, m^2 \, g^{-1}$. Zirconium titanate prepared by TON–ZON–TFTA process sintered at 1375°C shows a density of $4.32 \, g \, cm^{-3}$ (85% theoretical density). The SEM image of $Zr_5Ti_7O_{24}$ sintered at 1375°C (Fig. 8.22) shows compact sintered body with 98% theoretical density.

Fig. 8.22.   SEM of $ZrTiO_4$ sintered at 1375°C.

The dielectric loss of sintered zirconium titanates are in the order of 10^{-4}. The dielectric constants of $ZrTiO_4$ are in the range of 20–25 and that of $Zr_5Ti_7O_{24}$ are in the range of 31–35 at a frequency of 1 MHz. The sintered zirconium titanates shows low dielectric loss (2×10^{-4}) which is comparable with the value reported in literature.

8.8 ZrO_2–Ln_2O_3 SYSTEM: PYROCHLORES

Rare earth metal zirconates or pyrochlores $(Ln_2Zr_2O_7)$, Ln = La, Pr, Sm, Dy) are prepared by rapidly heating (350°C) an aqueous solution containing calculated amounts of zirconium nitrate, rare-earth-metal nitrate, and carbohydrazide/urea.[16]

The formation of $Ln_2Zr_2O_7$ with the pyrochlore structure was confirmed by powder XRD (Fig. 8.23). Often it is difficult to distinguish the pyrochlore and fluorite structures by XRD, because of the low intensity of pyrochlore (111), (331), and (511) XRD reflections. The particulate properties of combustion-derived rare earth zirconates are summarized in Table 8.14.

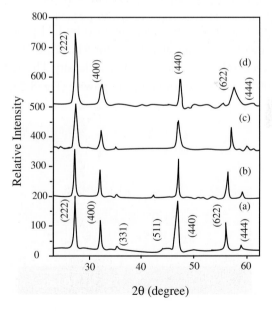

Fig. 8.23. XRD patterns of (a) $La_2Zr_2O_7$, (b) $Pr_2Zr_2O_7$, (c) $Sm_2Zr_2O_7$, and (d) $Dy_2Zr_2O_7$.

Table 8.14.   Particulate properties of rare earth zirconates.

Compound	Lattice parameter 'a' (nm)	Powder density (g cm^{-3})	Surface area (m^2 g^{-1})	Particle size[a] (nm)
La$_2$Zr$_2$O$_7$	1.0774	4.1 (4.4)	14 (6)	100 (220)
Ce$_2$Zr$_2$O$_7$	1.0701	4.3 (4.5)	18 (8)	70 (160)
Pr$_2$Zr$_2$O$_7$	1.0658	4.4 (4.8)	20 (10)	60 (120)
Nd$_2$Zr$_2$O$_7$	1.0623	4.5 (4.9)	19 (10)	60 (120)
Sm$_2$Zr$_2$O$_7$	1.0575	5.0 (5.1)	20 (9)	60 (130)
Gd$_2$Zr$_2$O$_7$	1.0503	5.2 (5.5)	16 (9)	70 (120)
Dy$_2$Zr$_2$O$_7$	1.0437	5.2 (5.6)	17 (10)	60 (100)

[a] From surface area; values in the parenthesis correspond to rare-earth-metal zirconates obtained by urea process.

The powders derived from the CH processes have a lower powder density than those prepared by urea process, indicating the porous nature of the combustion residue formed by the CH process. The specific surface areas of Ln$_2$Zr$_2$O$_7$ prepared by the CH process are higher (14–20 m^2 g^{-1}) than those obtained by urea process (6–10 m^2/g). The Pr$_2$Zr$_2$O$_7$ powders prepared by CH and urea show individual equiaxial particles and are virtually free from agglomeration.

The cold-pressed Pr$_2$Zr$_2$O$_7$ pellet obtained by CH process was sintered to 99% of the theoretical value at 1500°C for 4 h. While the powder derived from urea achieved only 87% of the theoretical density. The SEM of Pr$_2$Zr$_2$O$_7$ sintered at 1500°C reveals controlled grain growth (Fig. 8.24). Almost all the

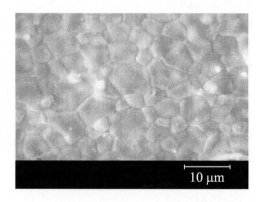

Fig. 8.24.   SEM of CH-derived Pr$_2$Zr$_2$O$_7$ sintered at 1500°C.

grains are well connected with linear grain boundaries and exhibit a nearly pore-free state of the sintered body.

8.9 NASICONs

NASICON ceramics have interesting properties such as low thermal expansion, high temperature stability, and fast ionic conductivity. They have great potential as hosts for the immobilization of radioactive wastes, catalyst supports, auto-engine components, heat exchangers, and mirror blanks for space technology. The superionic conductivity of NASICON has been widely investigated for use in energy storage and energy conversion systems along with microwave devices such as electrochemical sensors and electrochromic displays.

The structure of $NaZr_2P_3O_{12}$ (NZP), whose general formula is $Na_{1-x}Zr_2Si_xP_{3-x}O_{12}$, is composed of a three-dimensional skeletal network of PO_4 tetrahedra, corner sharing with ZrO_6 octahedra through strong Zr–O–P bonds. The most important feature in the structure is the stability and flexibility of its three-dimensional skeleton of $[Zr_2(PO_4)_3]^-$. The $[Zr_2(PO_4)_3]^-$ groups repeat along the threefold axis and the columns thus formed are connected in a hexagonal array. The Na^+ ions in $NaZr_2P_3O_{12}$ occupy the most stable octahedral interstitial sites (Fig. 8.25). Substitution of the interstitial cation by a larger cation (K or 1/2Ca) makes the structure expand in the c direction and contract in the a direction. The stable and flexible skeleton appears to resist these opposing variations in the lattice parameter. It also creates a dilatational stress against the two ZrO_6 octahedra which results in the rotation of PO_4 groups around the twofold axis. The structure of NZP consists of only framework atoms.

8.9.1 $MZr_2P_3O_{12}$ (M = Na, K, 1/2 Ca, and 1/4 Zr) and $NbZrP_3O_{12}$

The NASICON family of ceramics with the general formula $MZr_2P_3O_{12}$ (M = Na, K, 1/2 Ca and 1/4 Zr) and $NbZrP_3O_{12}$ are prepared by solution combustion method. In a typical experiment, an aqueous solution containing 1.88 g sodium nitrate, 5 g zirconyl nitrate, 1.028 g diammonium hydrogen

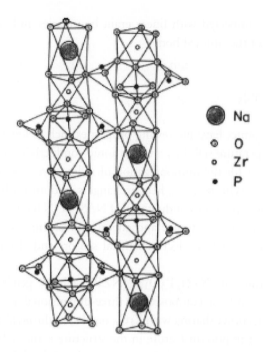

Fig. 8.25.   Structure of prototype NZP framework.

phosphate, 1.5 g ammonium perchlorate, and 4 g CH on heating rapidly at 400°C undergoes combustion to yield $NaZr_2P_3O_{12}$ powder.

The formation of $NaZr_2P_3O_{12}$ by solution combustion can be represented by the following reaction:

$$NaNO_3(aq) + 2ZrO(NO_3)_2(aq) + 3(NH_4)_2HPO_4(aq)$$
$$+ 5NH_4ClO_4(aq) + 3CH_6N_4O(aq)$$
$$\longrightarrow NaZr_2P_3O_{12}(s) + 3CO_2(g) + 30H_2O(g) + 14N_2(g)$$
$$+ 5HCl(g) + 2O_2(g)$$

$$\text{(54 mol of gases/mol of NASICON)} \quad (15)$$

The formation of NASICON by combustion takes place via a metathetically formed ammonium zirconium phosphate which undergoes deammoniation with metal ions.

The Nb compound is prepared by the combustion of an aqueous solution containing 2 g $NbOPO_4 \cdot 3H_2O$, 1.91 g zirconyl nitrate, 2.18 g diammonium hydrogen phosphate, 2.36 g ammonium perchlorate, and 1.5 g carbohydrazide. The $NbOPO_4 \cdot 3H_2O$ component is prepared by dissolving 2 g of metallic niobium in a mixture of an aqueous solution containing 10 ml of 40% HF and 5ml of concentrated HNO_3. Fifteen grams of 85% H_3PO_4 is added to this and the solution is heated in a water bath until a crystalline product precipitates. The precipitate is filtered off, washed with water and ethanol, and dried in air.[17]

The combustion-synthesized NASICONs are highly crystalline. However, $NbZrP_3O_{12}$ is weakly crystalline. Further calcinations of $NbZrP_3O_{12}$ at 600°C for 3 h yields a crystalline single phase powder (Fig. 8.26).

The lattice parameters of $NbZrP_3O_{12}$ are small compared with other NASICONs and $NbZrP_3O_{12}$ exhibits the highest surface area. NASICON formation has been further confirmed by their characteristic IR and ^{31}P MAS NMR spectra.[17] The particulate properties of combustion-derived NASICONs are summarized in Table 8.15.

Combustion-synthesized NASICONs sintered in the range of 1100–1200°C for 5 h achieve 85–90% of the theoretical density. The sintered density

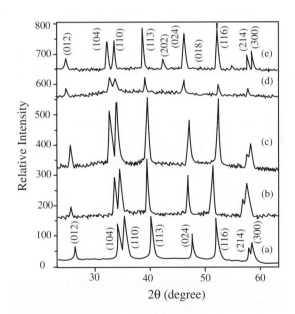

Fig. 8.26.   XRD patterns of $MZr_2P_3O_{12}$ (a) M = K, (b) M = 1/2 Ca, (c) M = 1/4 Zr, (d) $NbZrP_3O_{12}$, and (e) $NbZrP_3O_{12}$ calcined at 600°C.

Table 8.15.   Particulate properties of combustion-derived NASICONs.

Compound	Hexagonal		Powder density $(g\,cm^{-3})$	Surface area $(m^2\,g^{-1})$	Particle size[a] (nm)
	a (nm)	c (nm)			
$NaZr_2P_3O_{12}$	0.88	2.28	2.7	8	277
$KZr_2P_3O_{12}$	0.87	2.39	2.5	14	171
$Ca_{0.5}Zr_2P_3O_{12}$	0.88	2.28	2.4	15	166
$Zr_{0.25}Zr_2P_3O_{12}$	0.88	2.30	2.4	15	166
$NbZrP_3O_{12}$	0.87	0.23	2.4	28	89

[a] From surface area.

is further increased by the addition of 2 wt.% MgO. It is observed that the addition of MgO suppresses the grain growth as well as the microcracking and promotes the densification of the NASICON. Sintering of NASICONs above 1200°C decreases their density due to partial melting. The microstructure of $Ca_{0.5}Zr_2P_3O_{12}$ with 5 wt.% MgO sintered at 1200°C for 5 h shows the presence of dense thin needle-like grains (Fig. 8.27) and its EDAX analysis shows the presence of Mg- and P-rich phases in the thin-needle like grains. It is also observed that the addition of an excess amount of sintering aid like MgO leads to the formation of $Mg_3(PO_4)_2$ and t-ZrO_2 phases.

Sintered $NaZr_2P_3O_{12}$ shows the lowest linear thermal expansion coefficient of $-1.5 \times 10^{-6}\,K^{-1}$ and the largest linear thermal expansion coefficient

Fig. 8.27.   SEM of sintered $Ca_{0.5}Zr_2P_3O_{12}$ at 1200°C with 5 mass% of MgO.

is shown by $KZr_2P_3O_{12}$ (1×10^{-6} K^{-1}). The unusual low or negative thermal expansion coefficients of sintered polycrystalline NASICONs is due to the presence of microcracks. The large anisotropy in the thermal expansion of NASICON materials is due to the coupling of ZrO_6 octahedra and PO_4 tetrahedra causing constrained rotation during heating. About 5 wt.% MgO containing CZP exhibited near-zero thermal expansion coefficient, which is attributed to the presence of secondary minor phases such as magnesium phosphate and ZrO_2.

8.9.2 NASICON (Na Superionic Conductor) Materials ($Na_{1+x}Zr_2P_{3-x}Si_xO_{12}$)

NASICON, $Na_{1+x}Zr_2P_{3-x}Si_xO_{12}$ is prepared by the substitution of Si^{4+} for P^{5+} without changing the crystal framework of NZP (Fig. 8.25); the Na^+ used for charge compensation occupies the channels giving rise to fast-ion conductivity.

$Na_{1+x}Zr_2P_{3-x}Si_xO_{12}$ materials (where $x = 0, 0.05, 1.0, 1.5, 2,$ and 2.5) are prepared by the combustion of aqueous solutions containing stoichiometric amounts of sodium nitrate, zirconyl nitrate, fumed silica (surface area 100 m^2 g^{-1}), diammonium hydrogen phosphate, ammonium perchlorate, and carbohydrazide. When this redox mixture is rapidly heated at 400°C, it boils and ignites to burn with a flame to yield NASICON powders in less than 5 min.

In a typical experiment, 1.88 g sodium nitrate, 5 g zirconyl nitrate, 1.028 g diammonium hydrogen phosphate, 0.935 g fumed silica, 1.5 g ammonium perchlorate, and 4 g carbohydrazide are dissolved by stirring in 50 ml of water. Initially, the solution is heated on a hot plate to evaporate the excess water; then the resulting slurry is rapidly heated at 400°C. It undergoes decomposition with evolution of a larger amount of gas and ignites to burn with a flame, yielding voluminous $Na_3Zr_2PSi_2O_{12}$ powder. Ammonium perchlorate is used to oxidize diammonium hydrogen phosphate in the redox mixture.[18] Formation of $NaZr_2P_3O_{12}$ phosphates by solution combustion may be represented by the following reaction:

$$NaZr_2P_3O_{12} + xSiO_2 \longrightarrow Na_{1+x}Zr_2P_{3-x}Si_xO_{12}$$
$$(x = 0, 0.05, 1.0, 1.5, 2, \text{ and } 2.5) \qquad (16)$$

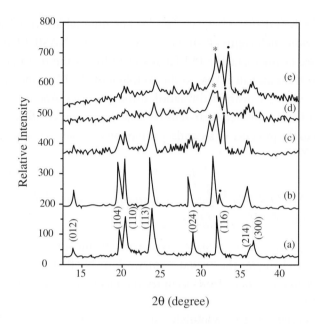

Fig. 8.28.   XRD patterns of as-prepared $Na_{1+x}Zr_2P_{3-x}Si_xO_{12}$: (a) 0.0, (b) 0.5, (c) 1.0, (d) 1.5, and (e) 2.5. ($\bullet$) $NaPO_3$, (*) t-ZrO_2.

Fumed SiO_2 present in the redox mixture reacts with the metathetically formed NZP (Eq. 15) to give $Na_{1+x}Zr_2P_{3-x}Si_xO_{12}$. The addition of energetic oxidizer like NH_4ClO_4 (Eq. 15) enhances the exothermicity of the combustion reaction to form NZP and NASICONs in the presence of ammonium phosphate and silica, respectively, which are well-known fire retardants.

The powder XRD patterns of combustion derived $Na_{1+x}Zr_2P_{3-x}Si_xO_{12}$ (x = 0–2.5) materials are shown in Fig. 8.28. The lattice parameters ($a = b = 0.88045$ nm and $c = 2.8$ nm) refined with least-squares fit using XRD reflections, are in good agreement with the literature. When fumed silica is added to the redox mixture, the crystallinity of NASICON powders is found to decrease and impurity phases such as tetragonal zirconia and $NaPO_3$ begin to appear. This is probably because fumed SiO_2 reduces the flame temperature in the redox mixture. Consequently, the crystallinity of NASICON materials also decreases with the gradual increase in SiO_2 content. Calcination of $Na_{1+x}Zr_2P_{3-x}Si_xO_{12}$ powders between 600°C and 800°C for 3 h yields crystalline NASICON materials. The XRD pattern of $Na_3Zr_2PSi_2O_{12}$ heated at 600°C shows the formation of crystalline product with a small amount of impurity phase. The intensity of the (104) reflection is less than that of the (110) reflection and the splitting is not well resolved. As the calcination temperature is increased to 700–800°C the impurity phases disappear and the intensity of (104) reflection increases compared to that of the (110) reflection. The increase in intensity of the (104) reflection implies the formation of a single-phase $Na_3Zr_2PSi_2O_{12}$ material with the NZP structure.

As-prepared $Na_3Zr_2PSi_2O_{12}$ powder does not show any SiO_2 phase in powder XRD pattern. Nevertheless, it is necessary to know the nature of SiO_2, i.e., whether Si occupies the P position in the NZP structure or if it exists as amorphous SiO_2. The ^{29}Si NMR spectrum of $Na_3Zr_2PSi_2O_{12}$ shows a broad resonance at -96.5 ppm. This is a low-field shift compared with the resonance frequency (-110 ppm) of pure SiO_2, and could be due to the substitution of Si for P in the NZP framework structures. The broad resonance of the as-prepared $Na_3Zr_2PSi_2O_{12}$ indicates short-range ordering of silicon (weakly crystalline). The ^{29}Si NMR of heat-treated $Na_3Zr_2PSi_2O_{12}$ at 700°C and 800°C shows a sharp resonance at -98 ppm. The sharpness of the spectrum indicates that long-range ordering occurs in the heat-treated $Na_3Zr_2PSi_2O_{12}$ material and confirms the substitution of P by Si in the framework structure of NZP.

The IR spectrum of $NaZr_2P_3O_{12}$ shows strong absorptions around $1030 \, cm^{-1}$ and 700-$600 \, cm^{-1}$ which are characteristic internal modes of PO_4 groups (T_d symmetry). A shoulder around $1000 \, cm^{-1}$ is due to the distortion in the PO_4 tetrahedra and the absorption around $360 \, cm^{-1}$ has been assigned to (Zr-O) vibrational frequency of ZrO_6 octahedra. As the Si content in NASICON increases, the intensity of the $1000 \, cm^{-1}$ peak also gradually increases. The IR spectra of NASICON indicate that structural disordering around PO_4 tetrahedra occurs with increasing silicon content.

The particulate properties of combustion-derived NASICON materials are summarized in Table 8.16.

The high surface area of NASICON materials obtained by the combustion process could be attributed to the large amount of cold gases evolved which dissipates heat. The $NaZr_2P_3O_{12}$ sintered compact at 1200°C shows microcracks and larger grains. The addition of 2 wt.% MgO suppressed the grain growth as well as microcracking and promoted the densification of the NASICON material. Sintering of NASICON above 1200°C leads to a decrease in the density indicating the formation of liquid phase.

Table 8.16.   Particulate properties of NASICON materials.

Compound	Lattice Parameters		Powder density $(g\,cm^{-3})$	Surface area $(m^2\,g^{-1})$	Particle size[a] (nm)
	a (nm)	c (nm)			
$NaZr_2P_3O_{12}$	0.88	2.28	2.7	8	277
$Na_{1.5}Zr_2P_{2.5}Si_{0.5}$	0.88	2.27	2.8	9	238
$Na_2Zr_2P_2SiO_{12}$	0.89	2.27	2.5	15	160
$Na_{2.5}Zr_2P_{1.5}Si_{1.5}O_{12}$	0.89	2.22	2.4	22	113
$Na_3Zr_2PSi_2O_{12}$[b]	1.56	9.21	2.4	28	89
$Na_{3.5}Zr_2P_{0.5}Si_{2.5}O_{12}$	0.91	2.26	2.5	30	80

[a] From surface area.
[b] Monoclinic — all the other compounds are rhombohedral.

The bulk thermal expansion coefficient of NASICON varies continuously from moderately negative to moderately positive values. The sintered NASICON ($NaZr_2P_3O_{12}$) body has the minimum bulk thermal expansion coefficient of $-3.4 \times 10^{-6}\,K^{-1}$ and $Na_{3.5}Zr_2P_{0.5}Si_{2.5}O_{12}$ has the maximum bulk thermal expansion coefficient of $4.1 \times 10^{-6}\,K^{-1}$. The large anisotropy in the thermal expansion of NZP materials is due to the coupling of the ZrO_6 octahedra and PO_4 tetrahedra causing constrained rotation during heating. The maximum conductivity of $0.236\Omega^{-1}\,cm^{-1}$ at 300°C was obtained for $Na_3Zr_2PSi_2O_{12}$ sintered at 1150°C for five hours.

8.10 CONCLUDING REMARKS

Nano-zirconia, stabilized zirconia, and zirconia composites with Al_2O_3, CeO_2, TiO_2, and Ln_2O_3 are easily synthesized by solution combustion approach. Transition and rare earth metal-ions-doped zirconia and zircon pigments also lend themselves to this approach. Carbohydrazide is an ideal fuel for combustion synthesis of zirconia and its composites. Complex mixed metal oxides like NASICONs (Na superionic conductors) can be prepared by a single heterogeneous solution combustion step. These nanopowders with large surface area are sinteractive to high density at low temperatures.

References

1. Stevens R, *Zirconia and Zirconia Ceramics*, Magnesium Elektron Ltd., Twickenham UK, 1986.

2. Arul Dhas N, Patil KC, Combustion synthesis and properties of tetragonal, monoclinic and partially and fully stabilized zirconia powders, *Int J Self-Propagating High-Temp Synth* **1**(4): 576–589, 1992.

3. Mimani T, Patil KC, Solution combustion synthesis of nanoscale oxides and their composites, *Mater Phys Mech* **4**: 134–137, 2001.

4. Ganesan R, Gnanasekaran T, Periaswami G, Srinivasa RS, A novel approach for synthesis of nanocrystalline yttria stabilized zirconia powder via polymeric precursor routes, *Trans Ind Ceram Soc* **64**: 149–156, 2005.

5. Arul Dhas N, Patil KC, Properties of magnesia stabilized zirconia powders prepared by combustion route, *J Mater Sci Lett* **12**: 1844–1847, 1993.

6. Shukla AK, Vandana Sharma, Arul Dhas N, Patil KC, Oxide-ion conductivity of calcia and yttria-stabilized zirconias prepared by a rapid combustion route, *Mater Sci Eng* **B40**: 153–157, 1996.

7. Arul Dhas N, Patil KC. Preparation and properties of tetragonal zirconia powders, *Int J Self-Propagating High-Temp Synth* **3**(4): 311–320, 1994.

8. Aruna ST, Muthuraman M, Patil KC, Synthesis and properties of Ni–YSZ cermet anode material for solid oxide fuel cells, *Solid State Ionics* **111**: 45–51, 1998.

9. Muthuraman M, Arul Dhas N, Patil KC, Preparation of zirconia-based color pigments by the combustion route, *J Mater Synth Process* **4**(2): 115–120, 1996.

10. Patil KC, Ghosh S, Aruna ST, Ekambaram S, Ceramic pigments: Solution combustion approach, *The Indian Potter* **34**: 1–9, 1996.

11. Kingsley JJ, Patil KC, Self-propagating combustion synthesis of tetragonal zirconia–alumina powders, *Ceram Trans* **12**: 217–224, 1990.

12. Arul Dhas N, Patil KC, Combustion synthesis and properties of zirconia–alumina powders, *Ceram Int* **20**: 57–66, 1994.

13. Gravie RC, The occurrence of metastable tetragonal zirconia as a crystalline effect, *J Phys Chem* **69**: 1238–1243, 1965.

14. Aruna ST, Patil KC, Combustion synthesis and properties of nanostructured Ceria-Zirconia solid solutions, *Nanostruct Mater* **10**(6): 955–964, 1998.

15. Muthuraman M, Patil KC, Studies on $ZrO_2:TiO_2$ system: Synthesis and properties of $ZrTiO_4$ and $Zr_5Ti_7O_{24}$, *Int J Self-propagating High-Temp Synth* **6**: 83–90, 1997.

16. Arul Dhas N, Patil KC, Combustion synthesis and properties of fine-particle rare earth metal zirconates, $Ln_2Zr_2O_7$, *J Mater Chem* **3**: 1289–1294, 1993.

17. Arul Dhas N, Patil KC, Combustion synthesis and properties of the NASICON family of materials, *J Mater Chem* **5**(9): 1463–1468, 1995.

18. Arul Dhas N, Patil KC, Controlled Combustion synthesis and properties of fine-particle Nasicon materials, *J Mater Chem* **4**(3): 491–497, 1994.

Chapter 9

Perovskite Oxide Materials

9.1 INTRODUCTION

Oxide materials with a general formula, ABO_3, are named after the mineral Perovskite ($CaTiO_3$). The crystal structure is a primitive cube, with the A-cation (alkali, alkaline earth, or rare earth ion) in the middle of the cube, the B-cation in the corner, and the anion (commonly oxygen), in the centre of the face edges (Fig. 9.1). The structure is stabilized by the six coordination of the B-cation (octahedron) and 12 of the A cation.[1] The chemical and physical properties of perovskite oxides vary with composition from dielectric to superconductivity and they find application as sensors, ferroelectrics, and catalysts. The uses of some of these perovskite oxides are summarized in Table 9.1.

9.2 DIELECTRIC MATERIALS

A dielectric material is a substance that is a poor conductor of electricity, but an efficient supporter of electrostatic fields. When the flow of current between opposite electric charged poles is kept to a minimum and the electrostatic lines of flux are not impeded or interrupted, an electrostatic field can store energy. This property is useful in capacitors, especially at radio frequencies. Dielectric oxide materials are classified as ferroelectrics, relaxor ferroelectrics, and microwave resonators on the basis of their properties like dielectric constant, dielectric loss, and piezoelectric coefficient as shown in Table 9.2.[2]

In the following sections, the preparation and properties of these materials synthesized by the solution combustion method are presented.

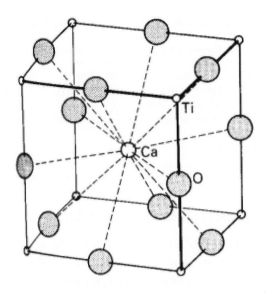

Fig. 9.1.   Perovskite structure ($CaTiO_3$).

Table 9.1.   Perovskite oxides and their applications.

Oxide	Properties	Applications
$MTiO_3$, M = Ca, Sr, Ba, Pb	Dielectric/pyroelectric/ Piezoelectric	Capacitors, acoustic transducer, sensors, pyrodetectors, photo anodes for solar energy conversion
$MZrO_3$, PZT, PLZT	Dielectric/pyroelectric/ piezoelectric/electro-optic	Gas lighters, pyroelectric sensors, wave guide device, optical memory display
PFN, PMN, PNN, PZN (Niobates)	Piezoelectric/electro- optic/dielectric/ electro-strictive	Relaxor ferroelectrics, second harmonic generation optical modulator, actuators
$LnMO_3$, M = Fe, Co, Cr, Mn	Antiferromagnetic	Heating element, Memory devices, SOFC materials — cathode and interconnect
$Ln(Sr)MO_3$, M = Mn, Co	Ferromagnetic	SOFC materials — cathode, CMR materials
$LaNiO_3$, $LaCuO_3$	Metallic	Metallic connectors for perovskites

Table 9.2.   Classification of dielectric materials and their properties.

Property	Ferroelectrics	Relaxor ferroelectrics	Microwave resonators
Dielectric constant (K)	100–1000	4000–40000	25–90
Dielectric loss	Low (<0.01)	Medium (<0.1)	Very low (<0.001)
Quality factor	Medium (>100)	Low (>10)	Very high (>1000)
Piezoelectric coefficient (pC/N)	High (>200)	Very high (>400)	Very low (>10)
Electrostatic coefficient (mm/kV)	Low	High	—

9.2.1 $MTiO_3$, $MZrO_3$ ($M = Ca$, Sr, and Ba)

Fine particles of $MTiO_3$, $MZrO_3$ (where M = Ca, Sr, and Ba) have been prepared by solution combustion using corresponding metal nitrates and tetraformal trisazine (TFTA).[3] For example, $BaTiO_3$ is prepared by the combustion of an aqueous redox mixture containing $Ba(NO_3)_2$, $TiO(NO_3)_2$, and TFTA in the mole ratio of 1:1:0.714. The combustion reaction is carried out at 350°C. On ignition, the redox mixture burns with a flame (temperature >1000°C) to yield crystalline, voluminous $BaTiO_3$. In general, assuming complete combustion the theoretical equation for the formation of $MTiO_3$ and $MZrO_3$ can be written as follows:

$$M(NO_3)_2(aq) + TiO(NO_3)_2(aq) + \underset{\text{TFTA}}{C_4H_{16}N_6O_2}(aq) + 2HNO_3(aq)$$
$$\longrightarrow MTiO_3(s) + 4CO_2(g) + 9H_2O(g) + 6N_2(g) + 1/2\, O_2(g)$$
$$(\sim 20\,\text{mol of gases/mol of } MTiO_3) \quad (1)$$

$$M(NO_3)_2(aq) + Zr(NO_3)_4(aq) + \underset{\text{TFTA}}{C_4H_{16}N_6O_2}(aq) + 2HNO_3(aq)$$
$$\longrightarrow MZrO_3(s) + 4CO_2(g) + 9H_2O(g) + 7N_2(g) + 3.5O_2(g)$$
$$(\sim 24\,\text{mol of gases/mol of } MZrO_3) \quad (2)$$

(M = Ca, Sr, and Ba)

Nitric acid is the excess oxidizer present in the mixture due to *in situ* generation of titanyl nitrate by the addition of nitric acid to titanyl hydroxide. It is well known that TFTA is hypergolic (a fuel which ignites in contact with an oxidizer) with fuming nitric acid and helps in lowering the ignition temperature.

The composition of the redox mixtures used for the combustion synthesis of metal titanates and metal zirconates and their particulate properties along with their dielectric properties are listed in Table 9.3.

Formation of $MTiO_3$ and $MZrO_3$ phases are confirmed by powder XRD patterns (Fig. 9.2). As-formed combustion products show the presence of $BaCO_3$ and TiO_2 impurities, which on calcination at $700°C$ for 30 min, yield pure perovskite phase.

The BET surface area of metal titanates and zirconates are in the range of 12–31 $m^2 g^{-1}$. The fine particle nature of $BaTiO_3$ confirmed by TEM image is 40 nm (Fig. 9.3). The $BaTiO_3$ pellet sintered at $1200°C$ for 2 h achieved 94–96% of theoretical density and showed uniform grain size of 3 μm (Fig. 9.4). The dielectric constant of $BaTiO_3$ is the highest (2018) and this is comparable with the sol–gel derived samples. The dielectric loss of these samples is very low (0.01–0.02).

Barium titanate ($BaTiO_3$) has also been prepared by the solution combustion method using three different barium precursors (BaO_2), $Ba(NO_3)_2$, and $Ba(CH_3COO)_2$, and fuels such as carbohydrazide (CH), glycine (GLY), and citric acid (CA) in the presence of titanyl nitrate.[4] Figure 9.5 shows the SEM images of these powders that indicate a porous microstructure of $BaTiO_3$. Scanning electron micrographic studies indicate the variation of pore size from mesopores (≥ 10 nm) to macropores depending upon the precursors and fuel. For example, $BaTiO_3$ obtained by the reaction involving citric acid shows very high-order porosity (Fig. 9.5c) with the pore sizes varying from 0.05 to 2 μm. The porous nature of $BaTiO_3$ samples is also revealed by the low powder densities of 1.5–4.0 $g cm^{-3}$ as well as high surface area ranging from 14 to 25 $m^2 g^{-1}$.

9.2.2 *Lead-Based Dielectric Materials (PbTiO₃, PbZrO₃, PZT, and PLZT)*

Lead-based perovskites, having the general formula $Pb(Zr,Ti)O_3$, encompass a large family of materials for applications as multilayer capacitors, piezoelectric sensors, transducers, electrostrictive actuators, pyroelectric detectors, and optical devices. Recently, electron emission from cold solids like lead zirconium titanate (PZT) and lanthanum-doped PZT (PLZT) have attracted much attention because of their use in fast microtriodes, flat panel displays,

Table 9.3.　Compositions of redox mixtures, particulate, and dielectric properties of $MTiO_3$ and $MZrO_3$.

Sample	Composition of redox mixtures	Powder density $(g\,cm^{-3})$	Surface area $(m^2\,g^{-1})$	Particle size[a] (nm)	Dielectric constant[b]	Dielectric loss[b]
$CaTiO_3$	4.37 g $Ca(NO_3)_2$ + 5 g $TiO(NO_3)_2$ + 5.14 g TFTA	2.60	20	120	152	0.02
$SrTiO_3$	5.63 g $Sr(NO_3)_2$ + 5 g $TiO(NO_3)_2$ + 5.14 g TFTA	2.67	31	70	470	0.01
$BaTiO_3$	6.95 g $Ba(NO_3)_2$ + 5 g $TiO(NO_3)_2$ + 5.14 g TFTA	4.60	30	40	2018	0.02
$CaZrO_3$	1.10 g $Ca(NO_3)_2$ + 2.89 g $Zr(NO_3)_4\cdot5H_2O$ + 1.30 g TFTA	4.31	12	110	—	—
$SrZrO_3$	1.43 g $Sr(NO_3)_2$ + 2.89 g $Zr(NO_3)_4\cdot5H_2O$ + 1.30 g TFTA	4.10	23	50	—	—
$BaZrO_3$	1.76 g $Ba(NO_3)_2$ + 2.89 g $Zr(NO_3)_4\cdot5H_2O$ + 1.30 g TFTA	4.40	18	70	—	—

[a] From surface area.
[b] Room temperature at 10 kHz.

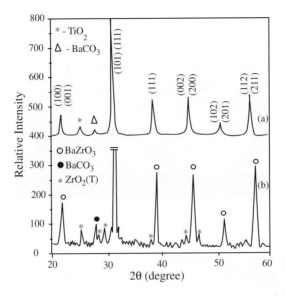

Fig. 9.2. XRD patterns of as-prepared (a) BaTiO$_3$ and (b)BaZrO$_3$.

Fig. 9.3. TEM of BaTiO$_3$ obtained from combustion method.

and bright electron guns. The electro-optic property of PLZT is exploited for use in optical shutters and switching. Lead-based dielectric materials (e.g., PbTiO$_3$, PLZT) are usually prepared by conventional ceramic methods. However, electronic ceramics prepared by such methods are sensitive to functions of compositional fluctuations near morphotropic phase, which is detrimental to the dielectric property of the perovskites.

Fig. 9.4.   SEM of $BaTiO_3$ sintered at 1200°C for 1 h.

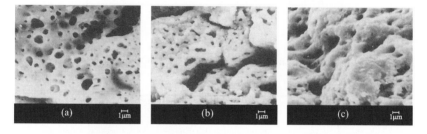

Fig. 9.5.   SEM of $BaTiO_3$ prepared by (a) CH, (b) citric acid, and (c) glycine.

Fine-particle dielectric oxide materials, $PbTiO_3$, $Pb(Zr_{0.53}Ti_{0.47})O_3$ (PZT), and $Pb_{0.92}La_{0.08}(Zr_{0.65}Ti_{0.35})_{0.98}O_3$ (PLZT) are prepared by the combustion of an aqueous solution, containing stoichiometric amounts of the corresponding metal nitrates and tetraformyl trisazine (TFTA) mixture, heated at 350°C.[3,5] The actual weights of the components in the redox mixture used for the combustion reaction are given in Table 9.4. The theoretical equation for the formation of $PbTiO_3$ assuming complete combustion can be written as follows:

$$Pb(NO_3)_2(aq) + TiO(NO_3)_2(aq) + \underset{TFTA}{C_4H_{16}N_6O_2}(aq) + 2HNO_3(aq)$$

$$\xrightarrow{350°C} PbTiO_3(s) + 4CO_2(g) + 9H_2O(g) + 6N_2(g) + 1/2O_2$$

$$(\sim 20 \text{ mol of gases/mol of } MTiO_3) \quad (3)$$

Similar equations can be written for $PbZrO_3$, PZT, and PLZT.

Table 9.4. Compositions of redox mixtures and particulate properties of lead-based dielectric materials.

Composition of redox mixtures	Pb-based dielectric materials	Perovskite %	Lattice constants (nm)			Surface area ($m^2 \, g^{-1}$)	Particle size[a] (nm)
			a	b	c		
$Pb(NO_3)_2$ 8.81 g + $TiO(N_3)_2$ 5.00 g + TFTA 5.14 g	$PbTiO_3$	95	0.39	0.39	0.48	30	30
$Pb(NO_3)_2$ 2.23 g + $Zr(NO_3)_4 \cdot 5H_2O$ 2.89 g + TFTA 1.30 g	$PbZrO_3$	95	0.47	0.47	0.48	13	60
$Pb(NO_3)_2$ 9.35 g + $TiO(NO_3)_2$ 2.49 g + $Zr(NO_3)_4 \cdot 5H_2O$ 6.43 g + TFTA 5.55 g	PZT	100	0.39	0.39	0.40	19	50
$Pb(NO_3)_2$ 6.61 g + $TiO(NO_3)_2$ 1.40 g + $Zr(NO_3)_4 \cdot 5H_2O$ 5.57 g + $La(NO_3)_3 \cdot 6 \, H_2O$ 0.15 g + TFTA 2.87 g	PLZT	100	0.57	0.59	0.41	30	20

[a] From surface area.

Formation of $PbTiO_3$, $PbZrO_3$, PZT, and PLZT has been confirmed from the powder XRD patterns (Fig. 9.6). It is well known that the preparation of pure perovskite phases is difficult as it is always accompanied by the pyrochlore phase.[6] The percentage of perovskite phase in the product varies from 95% to 100% as calculated from the formula given below:

$$\% \ \text{Perovskite} = \frac{I_{110\text{Perovskite}} \times 100}{I_{110\text{Perovskite}} + I_{222\text{Pyrochlore}}} \qquad (4)$$

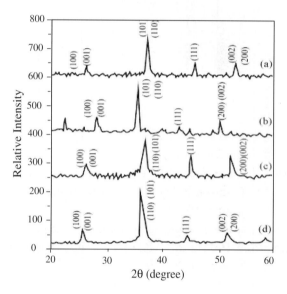

Fig. 9.6. Powder XRD patterns of as-prepared (a) $PbTiO_3$, (b) $PbZrO_3$, (c) PZT, and (d) PLZT.

Owing to the instability of PbO at elevated temperatures, trace amounts of the pyrochlore phase is also seen along with perovskite structure in the XRD patterns of $PbTiO_3$ and $PbZrO_3$.

The powder XRD patterns of PZT and PLZT show considerable line broadening indicating the nanoparticle nature of the products. As seen from the TEM image of PZT the particles are almost spherical in shape and are of 20 nm size (Fig. 9.7).

The particulate properties of $PbTiO_3$, PZT, and PLZT are summarized in Table 9.4. The trend in the surface area values of PZT and PLZT indicates solid solution formation. Solution combustion synthesized $PbTiO_3$, PZT, and

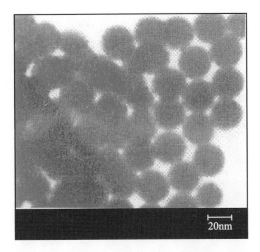

Fig. 9.7.   TEM of as-prepared PZT powder.

PLZT powders with polyvinyl alcohol (PVA) binder, when compacted and sintered at 1150°C for 2 h achieve 95–98% of theoretical density. The SEM images of the polished surface of $PbTiO_3$, PZT, and PLZT pellets sintered at 1150°C for 2 h are shown in Fig. 9.8. They show the presence of uniform grain size in the range of 5–10 μm.

The plot of variation of room temperature dielectric constant with frequency ranging from 100 Hz to 1 MHz for $PbTiO_3$ shows a large dielectric constant and a much lower dissipation factor. The dielectric constant and loss variation of PZT pellets between $-50°C$ and $200°C$ is plotted in Fig. 9.9. The dielectric properties of lead-based materials such as room temperature dielectric constants (K), dielectric loss ($\tan \delta$), and piezoelectric coefficient (d_{33}) of $PbTiO_3$ and PZT pellets are given in Table 9.5. An enormous increase in d_{33} values for PZT (335 pC/N) is seen after poling at 50 kV cm^{-1} in the temperature range of 110–120°C.

When compared with the expected range of the ferroelectric materials (Table 9.2), the observed dielectric properties of PZT fall within the range.

9.3   RELAXOR MATERIALS (PFN, PMN, PNN, AND PZN)

Lead-based relaxor ferroelectrics are increasingly used as capacitors, actuators, and piezoelectrics. The major challenge in manufacturing these relaxors is

Fig. 9.8.   SEM of the sintered pellet surface of (i) PbTiO$_3$, (ii) PZT, and (iii) PLZT.

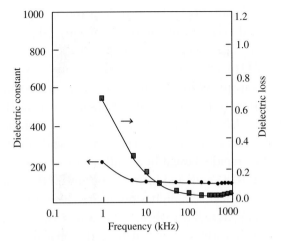

Fig. 9.9.   Variation of dielectric constant and loss versus frequency of redox mixture derived $PbTiO_3$ pellets.

Table 9.5.   Dielectric properties of lead-based dielectric materials.

Pb-based dielectric materials	Dielectric constant K at 25°C	Dielectric loss, tan δ at 25°C	Piezoelectric coefficient, d_{33} at 25°C (pC/N)
$PbTiO_3$	234	0.02	45
PZT	1024	0.03	335
PLZT	743	0.04	—

in the production of the materials with reliable and reproducible properties. Many of the important relaxors such as $Pb(Mg_{1/3}Nb_{2/3})O_3$[PMN] belong to the perovskite structure family. However, it is difficult to prepare pure perovskite phase (pyrochlore free)[6] relaxors by the conventional solid-state method. Lead-based relaxors with perovskite structure, free from undesirable pyrochlore phase have been obtained by the columbite precursor method. In this method, a two-stage calcinations step is followed: firstly, MgO is pre-reacted with Nb_2O_5 to form $MgNbO_6$ precursor which has columbite structure. Secondly, the $MgNbO_6$ is then reacted with PbO to obtain PMN with perovskite structure. Even though the formation of pyrochlore phase is suppressed by this method other problems such as PbO loss and variation in dielectric properties due to processing variables cannot be avoided.

Nanosize ferroelectrics like lead niobates lead iron niobate ($Pb(Fe_{1/2}Nb_{1/2})$$O_3$ — PFN), lead magnesium niobate ($Pb(Mg_{1/3}Nb_{2/3})O_3$ — PMN), lead nickel niobate ($Pb(Ni_{1/3}Nb_{2/3})O_3$ — PNN), and lead zinc niobate ($Pb(Zn_{1/3}Nb_{2/3})O_3$ — PZN) and their solid solutions with $BaTiO_3$ (BT) and $PbTiO_3$(PT) are prepared by solution combustion method.

Aqueous solutions containing calculated amounts of metal (Fe/Mg/Ni/Zn) nitrate, $Pb(NO_3)_2$, $Nb_2(C_2O_4)_5$, NH_4NO_3, and TFTA dissolved in a minimum amount of water are used for the combustion synthesis of lead niobates. The mixture when rapidly heated in a muffle furnace maintained at a temperature of $350 \pm 10°C$ boils and ignites to burn with the evolution of large amounts of gases to yield voluminous products. The as-formed combustion products are weakly crystalline; when calcined in a closed crucible at various temperatures (500–1000°C, for 30 min), they give fully crystalline lead niobates.[7]

The XRD patterns of the as-prepared and calcined powders of PMN and PZN are shown in Figs. 9.10(A) and 9.10(B). The XRD patterns show the presence of mixed phases such as perovskite, pyrochlore, $Pb_3Nb_2O_8$, and PbO

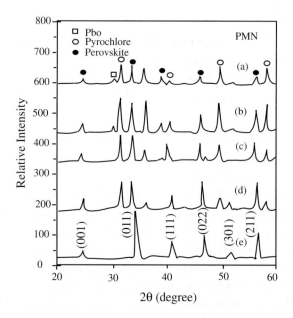

Fig. 9.10(A).   XRD patterns of PMN (a) as-prepared, (b) calcined at 500°C, (c) calcined at 600°C, (d) calcined at 700°C, and (e) calcined at 800°C.

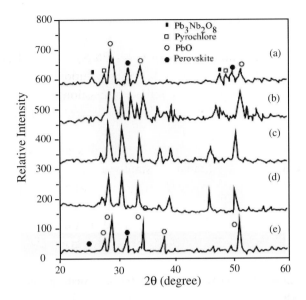

Fig. 9.10(B).   XRD patterns of PZN (a) as-prepared, (b) calcined at 500°C, (c) calcined at 600°C, (d) calcined at 700°C, and (e) calcined at 800°C.

or $Pb(NO_3)_2$. The percentage perovskite formed is in the range of 30–55 probably due to the short combustion time. However, the powders on calcination at 800°C for 30 min are converted into pure perovskite. The pure single-phase perovskite can be stabilized by forming solid solutions of PMN with PT and PZN with PT and BT. Surprisingly, PZN shows 60% perovskite at 600°C and at higher calcination temperatures the percentage of pervoskite decreases to 10% probably due to its thermodynamic instability. The powder XRD pattern of as-prepared lead nickel niobate (PNN) powders shows a mixture of perovskite, pyrochlore, and some unreacted PbO. On calcination at a temperature of 800°C for 2 h these samples show pure perovskite phase. The TEM image of as-formed particles of PZN–PT are fine and foamy with large pores and voids (Fig. 9.11), probably due to the large amount of gases evolved and low temperature of preparation. The particulate and dielectric properties of the combustion-derived relaxor powders are summarized in Table 9.6.

All these powders when sintered at 1050°C achieve 94–96% of theoretical density. Figure 9.12 shows the variation of the dielectric constant (K) and the dielectric loss (tan δ) with frequency and temperature for PMN and PMN–PT (90–10). As the amount of PT increases, both T_c and K value increase.

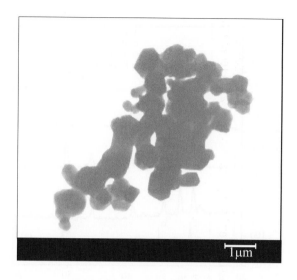

Fig. 9.11.   TEM of as-prepared PZN–PT (90/10) powder.

The value of T_c increases from $-8°C$ to $60°C$, whereas the value of K increases from 16,000 to 24,000. This occurs due to the formation of solid solution whose composition is similar to that of the morphotropic phase boundary. However, as the amount of PT increases, the relaxation in the dielectric constant with respect to frequency is reduced. This can be explained in terms of the change in their properties from relaxor ferroelectric to normal ferroelectric because of phase change.

9.4 MICROWAVE RESONATOR MATERIALS

The compounds, $A_2Nb_2O_7$ (A = Ca, Sr) and $Ln_2Ti_2O_7$ (Ln = La, Nd and Sm), are of interest for applications in high-frequency (GHz range) devices, owing to their ferroelectric properties and thermal resistance. Complex perovskite compounds, $Ba(B'_{1/3}B''_{2/3})O_3$ (B' = Mg, Zn and B'' = Nb, Ta), are reported to possess excellent microwave properties. These ceramics have to fulfill the requirements of high permittivity, extremely low dielectric loss, and low temperature coefficient of permittivity to yield temperature stable resonators.

Microwave resonator materials are prepared by the combustion of an aqueous redox mixture containing corresponding metal nitrates, niobium oxalate dissolved in ammonium nitrate, and TFTA heated at $350°C$.[8,9] The reaction is of flaming type. The chemical reaction, assuming complete combustion, can

Table 9.6. Particulate and dielectric properties of relaxor materials.

Compound	Powder density ($g\,cm^{-3}$)	Surface ($m^2\,g^{-1}$)	Particle size[a] (nm)	Dielectric constant, K		Dielectric loss, $\tan\delta$		Piezoelectric coefficient, d_{33} ($\times 10^2$ pC/N)
				10^3 (kHz)	10^3 (MHz)	10^{-2} (kHz)	10^{-2} (MHz)	
$(Pb(Fe_{1/2}Nb_{1/2})O_3$ — PFN)	6.1	2.5	300	0.9	0.9	2.4	1.8	2.5
$Pb(Mg_{1/3}Nb_{2/3})O_3$ — PMN)	6.2	17.0	100	1.2	1.1	1.3	1.2	0.5
$(Pb(Ni_{1/3}Nb_{2/3})O_3$ — PNN)	6.5	3.0	300	0.4	0.4	1.3	1.0	1.3
PMN–BaTiO$_3$ (90–10)	6.4	13.2	100	1.3	1.3	1.9	1.8	2.1
$(Pb(Zn_{1/3}Nb_{2/3})O_3$ — PZN)–BaTiO$_3$ (90–10)	6.4	3.2	300	8.1	7.9	2.0	1.0	2.9
$(Pb(Zn_{1/3}Nb_{2/3})O_3$ — PZN)–PbTiO$_3$ (90–10)	6.2	3.3	300	1.0	0.8	0.6	2.3	2.0

[a] From surface area.

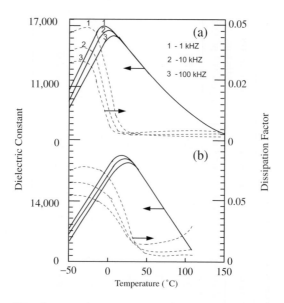

Fig. 9.12. Plot of K and tan δ with temperature and frequency of (a) PMN and (b) PMN–PT (90–10).

be written as

$$2Ln(NO_3)_3(aq) + 2TiO(NO_3)_2(aq) + 2C_4H_{16}N_6O_2(aq)$$
$$\text{TFTA}$$
$$\xrightarrow{3/2O_2} Ln_2Ti_2O_7(s) + 11N_2(g) + 8CO_2(g) + 16H_2O(g)$$
$$(35 \text{ mol of gases/mol of } Ln_2Ti_2O_7)$$
$$(Ln = La, \ Nd, \ and \ Sm) \hspace{3cm} (5)$$

$$2A(NO_3)_2(aq) + Nb_2(C_2O_4)_5(s) + 5NH_4NO_3(aq) + C_4H_{16}N_6O_2(aq)$$
$$\text{TFTA}$$
$$\xrightarrow{2O_2} A_2Nb_2O_7(s) + 14CO_2(g) + 18H_2(g) + 10N_2(g)$$
$$(42 \text{ mol of gases/mol of } A_2Nb_2O_7)$$
$$(A = Ca \ and \ Sr) \hspace{4cm} (6)$$

The particulate properties of the solution combustion synthesized microwave resonator materials are given in Table 9.7.

Figure 9.13 shows the variation of dielectric constant (K) and dielectric loss (tan δ) of $La_2Ti_2O_7$ and $Nd_2Ti_2O_7$ with temperature and frequency.

Table 9.7.   Particulate properties of microwave resonators.

Compound	Powder density $(g\,cm^{-3})$	Surface area $(m^2\,g^{-1})$	Particle size[a] (nm)
$La_2Ti_2O_7$	4.79	12.95	100
$Nd_2Ti_2O_7$	4.39	10.30	130
$Sm_2Ti_2O_7$	5.17	15.92	70
$Ca_2Nb_2O_7$	3.12	18.50	90
$Sr_2Nb_2O_7$	4.11	14.70	150

[a] From surface area.

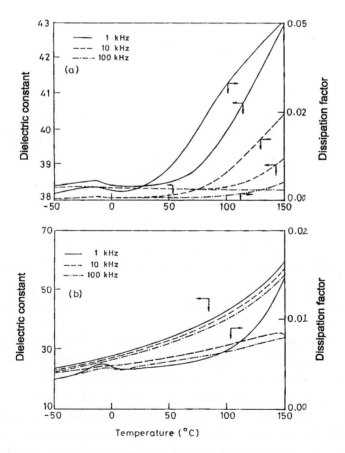

Fig. 9.13.   Variation of dielectric constant (K) and loss (tan δ) with temperature of sintered pellets of (a) $La_2Ti_2O_7$ and (b) $Nd_2Ti_2O_7$.

When the temperature increases from $-50°C$ to $150°C$ there is only a slight variation in the dielectric constant and dielectric loss. A significant change in dielectric constant and loss is observed as the frequency increases from 1 kHz to 100 kHz. Since there is no significant change in K with respect to temperature of $La_2Ti_2O_7$, the temperature coefficient of permittivity (TC_ε) is minimum (16 ppm K^{-1}). This is one of the important criteria for being a microwave resonator material. The variation of K and Q (quality factor) with frequency in the range of 100 Hz to 1 MHz is shown in Fig. 9.14. There is a peak at 0.9 MHz for $La_2Ti_2O_7$ with high quality factor at 20000, whereas for $Nd_2Ti_2O_7$ it is 3250 at 20 kHz. The low dielectric loss or high quality factor with even greater frequency dependence is possibly caused by absence of voids and higher densification.

The dielectric properties of microwave resonator ceramics are tabulated in Table 9.8.

Table 9.8.   Dielectric properties of microwave resonators.

Composition	Structure	Dielectric constant (K)	Quality factor (Q)	Linear expansion coefficient (α)	TC_ε (ppm/K) observed
$La_2Ti_2O_7$	Monoclinic	35	1006	9.9	16
$Nd_2Ti_2O_7$	Monoclinic	30	1492	10.2	245
$Sm_2Ti_2O_7$	Monoclinic	62	928	10.9	−480

Based on a simplified Clausius–Mosotti equation, the variation in permittivity of a dielectric with temperature can be related to the temperature variation of the dimensions of a material as $TC_\varepsilon = -\varepsilon\alpha$ (where linear expansion coefficient is α). The measured and calculated values of TC_ε are different from each other. The measured values show that the permittivity increases as the temperature increases, whereas the calculated values show that permittivity decreases with increasing temperature. If the simplified Clausius Mosotti equation is valid, this result suggests that the temperature coefficient of polarisability might be comparable with or considerably exceed the volume expansion coefficient. Interestingly, most of the dielectrics showing exceptional behavior can be classified as those promising for dielectric resonators in microwave circuitry. The niobates when pelletized and sintered at $1350°C$ for 2 h achieve 98% theoretical density.

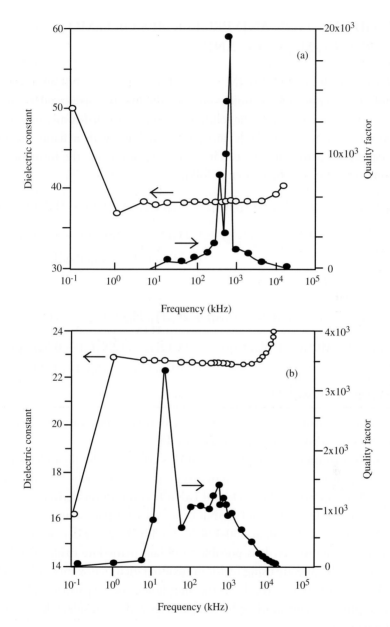

Fig. 9.14. Variation of dielectric constant (K) and quality factor (Q) with frequency of sintered pellets of (a) $La_2Ti_2O_7$ and (b) $Nd_2Ti_2O_7$.

9.5 PREPARATION AND PROPERTIES OF LnMO$_3$ (M = Cr, Mn, Fe, Co, AND Ni)

Perovskite oxides of LaMO$_3$ (M = Cr, Mn, Fe, Co, and Ni) are prepared by solution combustion of stoichiometric amounts of corresponding metal nitrates with TFTA fuels.[10] The solution containing the redox mixtures when rapidly heated at 350°C boils and ignites to burn with a flame yielding corresponding perovskite oxide. Theoretical equations for the formation of various perovskite can be written as follows:

$$2La(NO_3)_3(aq) + 2M(NO_3)_3(aq) + 2C_4H_{16}N_6O_2\ (aq) + 2HNO_3(aq)$$
$$\text{TFTA}$$
$$\longrightarrow 2LaMO_3(s) + 8CO_2(g) + 17H_2O(g) + 13N_2(g) + 3.5O_2(g)$$
$$(\sim 21\ \text{mol of gases/mol of LaMO}_3)$$

(M = Cr, Fe)   (7)

$$2La(NO_3)_3(aq) + 2M(NO_3)_2(aq) + 2C_4H_{16}N_6O_2\ (aq) + 2HNO_3(aq)$$
$$\text{TFTA}$$
$$\longrightarrow 2LaMO_3(s) + 8CO_2(g) + 17H_2O(g) + 12N_2(g) + 0.5O_2(g)$$
$$(\sim 19\ \text{mol of gases/mol of LaMO}_3)$$

(M = Mn, Co, and Ni)   (8)

The composition of the redox mixtures used for the combustion and particulate properties of LaMO$_3$ are summarized in Table 9.9

The formation of single-phase LaMO$_3$ is confirmed by their characteristic X-ray powder diffraction patterns. Typical XRD of LaCrO$_3$ is shown in Fig. 9.15. In the case of LaCoO$_3$ and LaNiO$_3$ combustion-derived powders, the presence of La$_2$CoO$_4$ and La$_2$NiO$_4$ is seen with trace amounts of CoO and NiO. On calcination at 600–800°C for 1 h these powders form single-phase perovskite oxides. It is possible to obtain perovskite oxides LaCoO$_3$, and LaNiO$_3$ by directly using fuel-lean (oxidizer-rich) redox mixtures.[10] The TEM of LaCrO$_3$ (Fig. 9.16) shows the morphology of the particles. The combustion process yields nearly spherical particles with a considerable degree of agglomeration.

Sintering study of fine-particle LaCrO$_3$ was carried out by making compacts of the powder by uniaxial pressing at 5 MPa. The cylindrical compacts (12 × 3 mm^2) when sintered at 1500°C for 5 h without adding sintering aids

Table 9.9.   Compositions of redox mixtures and particulate properties of LaMO$_3$ oxides.

Composition of the redox mixtures	Product	Density (g cm^{-3})	Surface area (m^2 g^{-1})	Particle size[a] (nm)
5.413 g La(NO$_3$)$_3$·9H$_2$O + 5.0 g Cr(NO$_3$)$_3$·9H$_2$O + 2.41 g TFTA	LaCrO$_3$	3.61	9	190
11.21 g La(NO$_3$)$_3$·9H$_2$O + 6.5 g Mn(NO$_3$)$_2$·4H$_2$O + 6.93 g TFTA	LaMnO$_3$	4.9	5	270
5.413 g La(NO$_3$)$_3$·9H$_2$O + 5.0 g Fe(NO$_3$)$_3$·9H$_2$O + 2.41 g TFTA	LaFeO$_3$	4.96	10	130
3.42 g La(NO$_3$)$_3$·9H$_2$O + 2.3 g Co(NO$_3$)$_2$·4H$_2$O + 2.841 g TFTA	LaCoO$_3$	4.12	13	370
3.42 g La(NO$_3$)$_3$·9H$_2$O + 2.29 g Ni(NO$_3$)$_2$·4H$_2$O + 2.841 g TFTA	LaNiO$_3$	3.95	4	410

[a] From surface area.

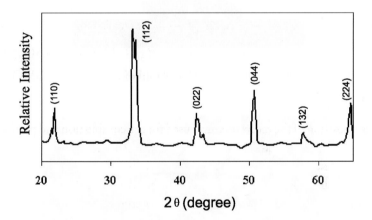

Fig. 9.15.   XRD pattern of LaCrO$_3$.

achieved only 85% theoretical density (5.2 g cm^{-3}). A SEM image of the sintered compact shown in Fig. 9.17 reveals uniform fine-grain growth demonstrating that a uniform particle size provides a measure of stability against exaggerated grain growth.

Rare earth chromites are of interest as interconnect material in solid oxide fuel cells (SOFC). The perovskite LaCrO$_3$ has high electrical conductivity and thermal coefficient of expansion similar to that of the eletrolyte.[11] Similarly,

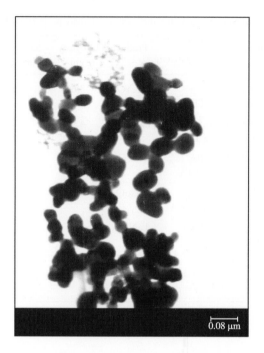

Fig. 9.16.   TEM of LaCrO$_3$.

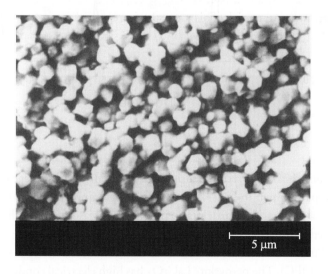

Fig. 9.17.   SEM of sintered LaCrO$_3$.

other rare earth chromites: $LnCrO_3$, (where $Ln = Pr, Nd, Sm, Gd, Dy,$ and Y) are prepared by the combustion of corresponding metal nitrates and TFTA mixtures. The composition and particulate properties of $LnCrO_3$ are summarized in Table 9.10.

Table 9.10. Compositions of redox mixtures and particulate properties of rare earth orthochromites.

Composition of Redox Mixtures	Product	Density $(g\,cm^{-3})$	Surface area $(m^2\,g^{-1})$	Particle size[a] (nm)
5.413 g $La(NO_3)_3 \cdot 9H_2O$ + B + C	$LaCrO_3$	3.61	8.7	190
6.111 g $Pr(NO_3)_3 \cdot 9H_2O$ + B + C	$PrCrO_3$	4.5	6.9	190
5.479 g $Nd(NO_3)_3 \cdot 9H_2O$ + B + C	$NdCrO_3$	4.8	8.0	150
5.555 g $Sm(NO_3)_3 \cdot 9H_2O$ + B + C	$SmCrO_3$	4.63	11.8	100
5.642 g $Gd(NO_3)_3 \cdot 9H_2O$ + B + C	$GdCrO_3$	4.8	15.5	80
5.482 g $Dy(NO_3)_3 \cdot 9H_2O$ + B + C	$DyCrO_3$	4.7	6.6	190
4.787 g $Y(NO_3)_3 \cdot 9H_2O$ + B + C	$YCrO_3$	4.0	16.9	80

B = 5.0018 g $Cr(NO_3)_3 \cdot 9H_2O$; C = 2.4135 g tetraformal trisazine.
[a] From surface area.

Since $LaCrO_3$ cannot be sintered to full density, calcium-and-cobalt substituted lanthanum chromite ($La_{0.7}Ca_{0.3}Cr_{0.9}Co_{0.1}O_3$) is being used as the interconnect material. This material is prepared by solution combustion method using stoichiometric amounts of lanthanum nitrate, chromium nitrate, and maleic hydrazide (MH) as fuel.[12] The combustion reaction is carried out at 350°C and is of flaming type yielding a voluminous product. The formation of $LaCrO_3$ and substituted $LaCrO_3$ can be represented by the following equations.

$$2La(NO_3)_3(aq) + 2Cr(NO_3)_3(aq) + \underset{MH}{3C_4H_8N_4O_2}(aq)$$
$$\longrightarrow 2LaCrO_3(s) + 12CO_2(g) + 12H_2O(g) + 12N_2(g)$$
$$(18 \text{ mol of gases/mol of } LaCrO_3) \qquad (9)$$

$$2La(NO_3)_3(aq) + 2Cr(NO_3)_3(aq) + \underset{MH}{3C_4H_8N_4O_2}(aq)$$
$$\xrightarrow{Ca^{2+}/Co^{2+}} 2LaCrO_3(s) + 12CO_2(g) + 12H_2O(g) + 12N_2(g)$$
$$(18 \text{ mol of gases/mol of } (LaCa)(CoCr)O_3) \quad (10)$$

The $La_{0.7}Ca_{0.3}Cr_{0.9}Co_{0.1}O_3$ powder is fine and crystalline. The nanosize nature of the as-formed powders (14 nm) is evident from the X-ray line broadening. X-ray pattern corresponds to orthorhombic phase and the lattice parameters are in agreement with the literature values. The interconnect powder has a high surface area of 34 $m^2\,g^{-1}$. Substituted $LaCrO_3$ can be sintered to >85% theoretical density at 1350°C for 5 h. The SEM image shows well-formed dense grains with a grain size of 0.5–2 μm (Fig. 9.18). It indicates a high conductivity value of 23 $S\,cm^{-1}$ at 900°C and the plot of log (σT) versus $1/T$ is linear, revealing thermally activated hopping of small polarons as the conduction mechanism. The thermal expansion is linear in the investigated temperature range and the thermal expansion coefficient is found to be $11.65 \times 10^{-6}\,K^{-1}$ suitable for SOFC application.

Fig. 9.18.   SEM of sintered $La_{0.7}Ca_{0.3}Cr_{0.9}Co_{0.1}O_3$.

A complete range of solid solutions between lanthanum manganite and lanthanum cobaltite are successfully prepared by the combustion process and examined as probable candidates for cathode material of SOFC.[13]

It is observed that substitution of Co into $LaMnO_3$ greatly influence the conductivity behavior. Combustion-derived $LaMnO_3$ shows high room-temperature electrical conductivity ($\sigma_{RT} = 6.3\,S\,cm^{-1}$) due to greater concentration of Mn^{4+}, while it shows a gradual decrease of electrical conductivity with increasing cobalt concentration. The decrease in conductivity at higher substitution of Mn may be due to the dilution of Co ions by Mn ions in the lattice. The conductivity value is highest for $LaCoO_3$ and decreases with increasing Mn content. The cubic $LaMn_{1-x}Co_xO_3$ ($x = 0.2$–0.5) shows lower conductivity values. In general, the conductivity values decrease with increasing unit cell volume. This decrease may be attributed to the increase in M–O bond length (M = Mn or Co), which, in turn, decreases the ease of an electron double exchange process.

The thermal expansion coefficient increases with increasing cobalt concentration. The solid solutions $LaMn_{0.7}Co_{0.3}O_3$ and $LaMn_{0.4}Co_{0.6}O_3$ have thermal expansion coefficient values of 12.5×10^{-6} and $16.3 \times 10^{-6}\,K^{-1}$, respectively which are higher than that of $LaMnO_3$ ($11.33 \times 10^{-6}\,K^{-1}$). Thus, only the compositions with very low Co content ($x < 0.2$), satisfy the conductivity and thermal expansion requirements for SOFC application.

9.6 PREPARATION AND PROPERTIES OF $La_{1-x}Sr_xMO_3$ (M = Mn AND Fe)

Both strontium-substituted rare earth manganite and ferrite are of interest as cathode material in SOFC, colossal magnetoresistance (CMR) materials, and electrode material.

Strontium-substituted rare earth manganites are prepared in a similar way described earlier (Eq. (8)) by the addition of required amount of strontium nitrate to the redox mixture containing lanthanum nitrate, manganese nitrate, and TFTA.[14] Strontium-substituted rare earth manganite $Ln_{1-x}Sr_xMnO_3$ (where Ln = Pr, Nd and Sm ; $x = 0, 0.16, 0.25$) can also be prepared by the solution combustion of an aqueous redox solution containing corresponding metal nitrates and ODH.[15] When ODH is used as fuel, the atmospheric oxygen oxidizes Mn^{2+} to Mn^{3+}/Mn^{4+} during the combustion.

Formation of a single-phase $La_{1-x}Sr_xMnO_3$ product is confirmed by powder XRD (Fig. 9.19). The lattice constants calculated from XRD match well with the reported values for $LaMnO_3$ ($a = 0.55$ nm, $b = 0.57$ nm, and $c = 0.77$ nm are equal to the calculated values). Analytical data and some properties of $La(Sr)MnO_3$ are summarized in Table 9.11

It may be noted that with increase in strontium content the Mn^{4+} concentration increases steadily. The presence of Mn^{3+}/Mn^{4+} couple leads to nonstoichiometry in these oxides. The powder density values of these oxides are in the range of 55–65% of the theoretical value. The specific surface area of the oxides is in the range of 5–15 $m^2\,g^{-1}$. It is interesting to note that

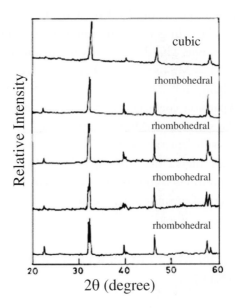

Fig. 9.19. XRD patterns of $La_{1-x}Sr_xMnO_3$ sintered at 1350°C, $x < 0.3$ rhombohedral & $x = 0.3$ cubic.

Table 9.11.   Analytical data and some properties of $La_{1-x}Sr_xMnO_3$.

x	Wt.% Mn(III)	Wt.% Mn(IV)	Total Mn% (theoretical value)	Density $(g\,cm^{-3})$	Surface area $(m^2\,g^{-1})$	Particle size[a] (nm)
0.0	19.86	2.84	22.70 (22.71)	4.90	4.5	270
0.2	19.45	4.23	23.69 (23.72)	4.70	7.5	170
0.3	20.00	4.30	24.30 (24.25)	4.75	9.5	130
0.4	20.50	4.35	24.85 (24.87)	4.68	12.0	100
0.5	19.75	5.59	25.34 (25.40)	4.50	14.5	90

[a] From surface area.

with an increase in strontium content in the redox mixture, the *in situ* flame temperature is lower compared to the unsubstituted $LaMnO_3$. As a result, $La_{0.5}Sr_{0.5}MnO_3$ has large surface area compared to $LaMnO_3$. The particle size $(0.2\,\mu m)$ as calculated from TEM image of $LaMnO_3$ (Fig. 9.20) shows that the particles are of uniform size and shape. The SEM image of $La_{1-x}Sr_xMnO_3$ (Fig. 9.21) shows large, dense, sintered "chunks" of the as-prepared powder, which is particularly useful for feeding through a plasma spray gun to form fuel-cell electrodes.

Fig. 9.20.   TEM of LaMnO$_3$.

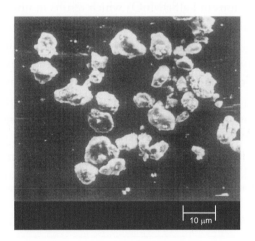

Fig. 9.21.   SEM of as-prepared La$_{1-x}$Sr$_x$MnO$_3$.

As-formed lanthanum manganites prepared from ODH fuel for SOFC application are found to be X-ray crystalline, corresponding to cubic symmetry. The strong oxidizing atmosphere which exists during combustion reaction of the redox mixture leads to excess Mn^{4+} in the LaMnO$_3$ structure stabilizing the cubic phase. Additionally, the quenching of the product from a flame temperature (*ca.* 900°C) to room temperature appears to favor the stabilization of Mn^{4+}. The Mn^{4+} content in the as-formed LaMnO$_3$ is 36% as estimated from iodiometry.

Calcination of as-formed $LaMnO_3$ at 700°C stabilizes the cubic phase. On further calcination at 1000°C, the Mn^{4+} content decreases to 28% and the phase changes to rhombohedral. As-formed $La_{1-x}Sr_xMnO_3$ also shows a cubic phase and the lanthanum manganites with $x < 0.3$ change to rhombohedral ($Mn^{4+} \approx 24\%$) after calcination at 1350°C for 2 h. $La_{1-x}Sr_xMnO_3$ with $x = 0.3$ when calcined at 1350°C shows a cubic phase with 44% Mn^{4+}. The stabilization of rhombohedral and cubic phases in $La(Sr)MnO_3$ at higher temperatures is due to the substitution of Sr^{2+} in La^{3+} sites resulting in higher Mn^{4+} content. The combustion-derived $LaMnO_3$ always contain >20% Mn^{4+} and the orthorhombic phase is not observed in any of the samples. The lattice parameter calculations from the XRD patterns of the sintered samples at 1350°C (Table 9.12) show that the rhombohedral unit-cell dimensions (a and α) decreases with increasing strontium substitution in $LaMnO_3$. This may be due to the increase in the concentration of smaller Mn^{4+} ions compared with Mn^{3+} ions in $La(Sr)MnO_3$ which results in unit cell contraction.

The microstructure of strontium-substituted lanthanum manganites sintered at 1350°C (Fig. 9.22) are suited for SOFC application as it shows a porous nature of the sintered body and interconnected pores of the fractured surface, as revealed in the SEM image. The thermal expansion coefficients, α (Table 9.12) increase with increasing substitution of strontium (Fig. 9.23). At lower substitution levels the thermal expansion coefficient (α) value is close to that of yttria-stabilized zirconia ($11.5 \times 10^{-6} \, K^{-1}$), the electrolyte of SOFC. At lower levels of Sr^{2+} ion substitution in lanthanum sites ($x \leq 0.1$), semiconductor like behavior is observed in the temperature range investigated,

Table 9.12.   Properties of sintered $La(Sr)MnO_3$ (1350°C, 4 h).

Composition	Phase (from XRD)	Lattice parameters		Thermal expansion coefficient at 900°C $(K^{-1} \times 10^{-6})$	Conductivity at 900°C $(S \, cm^{-1})$	Activation energy (eV)
		a (nm)	α (°)			
$LaMnO_3$	R	0.553	60.61	11.33	103	0.164
$La_{0.90}Sr_{0.10}MnO_3$	R	0.552	60.51	12.18	166	0.155
$La_{0.84}Sr_{0.16}MnO_3$	R	0.549	60.41	12.63	202	0.084
$La_{0.80}Sr_{0.20}MnO_3$	R	0.547	60.30	13.13	155	0.090
$La_{0.7}Sr_{0.3}MnO_3$	C	0.776	—	13.74	144	0.100

R = rhombohedral and C = cubic.

Fig. 9.22.   SEM of sintered $La_{1-x}Sr_xMnO_3$.

i.e., a decrease in resistivity with increasing temperature. At higher Sr^{2+} concentrations, $x \geq 0.16$, metallic behavior is observed at low temperatures, i.e., increase in resistivity with temperature. After a particular temperature, it behaves as a typical semiconductor and this transition temperature (T_t) coincides with the ferromagnetic T_c of the material.

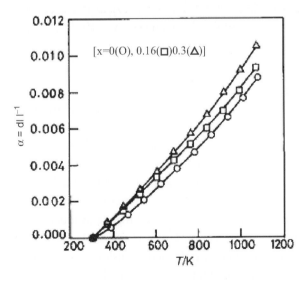

Fig. 9.23.   Plots of thermal expansion versus temperature for $La_{1-x}Sr_xMnO_3$.

The resistivity behavior below T_c is ascribed to the transport of holes in an extended band and above T_c to the hopping motion of localized holes. The plot of log (σT) versus $1/T$ of $La_{1-x}Sr_xMnO_3$ with $x < 0.16$ is linear in the measured temperature range and for $x \geq 0.16$ it is linear above the transition temperature (T_t). The linear dependence of log (σT) versus $1/T$ (Fig. 9.24) is characteristic of the polaron hopping transport mechanism for which conductivity can be represented by the function $\sigma = (A/T) \exp(-E_a/kT)$.

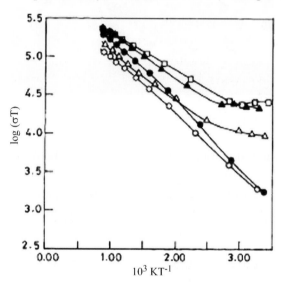

Fig. 9.24.   Plot of conductivity versus temperature for $La_{1-x}Sr_xMnO_3$.

The activation energy (E_a) for electrical conduction calculated from the slope of log (σT) versus $1/T$ decreases with increasing strontium substitution (Table 9.12). $La_{0.84}Sr_{0.16}MnO_3$ shows highest conductivity of 202 S cm^{-1} at 900°C in air for the same composition (190 S cm^{-1}) having maximum theoretical density. The higher conductivity may be due to a combination of surface and grain boundary effects as well as higher Mn^{4+} content. Also, the good contacts between grains and the interconnected pores in the sintered body may be responsible for the observed higher conductivity.

Strontium-doped lanthanum ferrite ($La_{1-x}Sr_xFeO_3$ where $x = 0–1$) is prepared by solution combustion method using stoichiometric amounts of metal nitrates and TFTA/ODH mixtures heated at 350°C.[15] The combustion reaction is smooth without a flame in case of ODH fuel, and flaming type in case of TFTA. Theoretical equation assuming complete combustion of the redox mixture can be written as follows:

$$La(NO_3)_3(aq) + Fe(NO_3)_3(aq) + 3C_2H_6N_4O_2(aq)$$
$$ODH$$

$$\xrightarrow{xSr(NO_3)_2} La_{1-x}Sr_xFeO_3(s) + 9CO_2(g) + 9H_2O(g) + 9N_2(g)$$
$$(27 \text{ mol of gases/mol of } La_{1-x}Sr_xFeO_3) \quad (11)$$

The formation of single-phase $La_{1-x}Sr_xFeO_3$ is confirmed by powder XRD study. The as-formed ferrites obtained by ODH method are found to be amorphous, which on heating to 650°C for 1 h give single-phase ferrites. On the other hand, in the TFTA method the as-formed material is found to be crystalline. The particulate properties of single-phase $La_{1-x}Sr_xFeO_3$ are summarized in Table 9.13.

Table 9.13. Particulate properties of $La_{1-x}Sr_xFeO_3$ perovskites.

Sample	Powder density (g cm^{-3})		Surface area (m^2 g^{-1})		Particle size[a] (nm)	
	ODH	TFTA	ODH	TFTA	ODH	TFTA
LaFeO$_3$	4.21	4.96	21	10	67	130
La$_{0.9}$Sr$_{0.1}$FeO$_3$	4.19	4.83	27	11	53	110
La$_{0.8}$Sr$_{0.2}$FeO$_3$	4.17	4.72	32	13	45	100
La$_{0.6}$Sr$_{0.4}$FeO$_3$	4.17	4.65	38	15	38	90
La$_{0.4}$Sr$_{0.6}$FeO$_3$	4.13	4.52	45	15	32	90
La$_{0.1}$Sr$_{0.9}$FeO$_3$	4.02	4.26	50	16	30	90
SrFeO$_{3-\delta}$	3.97	4.15	54	16	28	90

[a] From surface area.

The BET surface areas of the as-formed ferrites obtained from ODH process is two to three times higher compared to TFTA derived ferrites. The surface area increases linearly with strontium content. The typical plots of log resistivity versus $1/T$ compositions of $La_{1-x}Sr_xFeO_3$ are shown in Fig. 9.25. The data indicate semiconducting behavior. The oxygen evolution reaction (OER) and oxygen reduction reaction (ORR) experiments are carried out by sandwiching the sample between two degreased nickel wire mesh. The steady-state galvanostatic polarization plots for $La_{1-x}Sr_xFeO_3$ ($x \geq 0.8$) electrodes, for cathode as well as anodic modes are shown in Fig. 9.26; the kinetic parameters for the substituted and heat treated ferrites are given in Table 9.14.

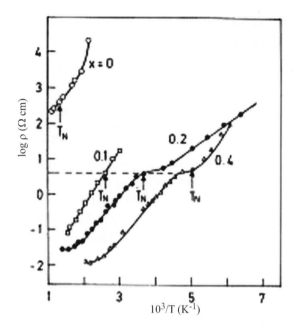

Fig. 9.25. Resistivity versus temperature curve for substituted orthoferrite $La_{1-x}Sr_xFeO_3$ ($x = 0$–0.4).

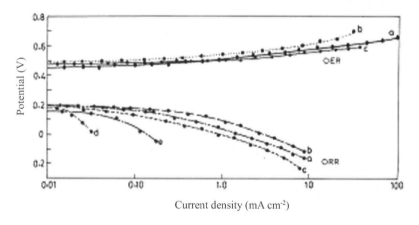

Fig. 9.26. Polarization curves in 6 M KOH at 30°C toward both OER and ORR (heat-treated samples): (a) $La_{0.8}Sr_{0.2}FeO_3$ (ODH method), (b) $La_{0.8}Sr_{0.2}FeO_3$ (TFTA method), (c) $SrFeO_3$ (ODH method), (d) $La_{0.8}Sr_{0.2}FeO_3$ (as-prepared TFTA method), and (e) $La_{0.8}Sr_{0.2}FeO_3$ (as-prepared ODH method).

Table 9.14.   Electrocatalytic activities of orthoferrites.

Sample	Tafel slope	OER (mV/decade)	Tafel slope	ORR (mV/decade)
	ODH	TFTA	ODH	TFTA
$La_{0.8}Sr_{0.2}FeO_3$	55	58	130	115
$La_{0.9}Sr_{0.1}FeO_3$	62	65	120	105
$SrFeO_{3-\delta}$	85	87	115	125

It is interesting to note that the Tafel slopes for these electrocatalysts lie between 60 and 90 mV/decade toward OER whereas ORR varies from 80 to 130 mV/decade. The as-formed powders are polarized at a very low current and the expected values are obtained only after heating at 1200–1250°C for 3 h followed by another treatment at 850°C in oxygen atmosphere for 24 h. The TFTA derived ferrites show a slightly better activity. The electrochemical activity of the ferrites is comparable to those reported in the literature. This indicates that perovskite phase formation is an important step for electrochemical activity. Similarly, a number of nanocrystalline $LnMnO_3$ (where Ln = Pr, Nd, and Sm); $La(M)MnO_3$ (where M = Ca, Sr, and Ba) perovskite materials have been prepared and investigated using solution combustion method.[16–19]

9.7 CONCLUDING REMARKS

Solution combustion method has been successfully used for preparing all types of technologically useful perovskite oxides ranging from dielectrics, ferroelectric materials, SOFC powders, and electrocatalytically active ferrites. TFTA appears to be an ideal fuel for the preparation of majority of perovskite oxides. The dielectric properties of combustion-synthesized perovskite oxides like $MTiO_3$, $MZrO_3$, and PZT are comparable with those prepared from sol–gel and solid-state methods.

References

1. Mc Carrol WH, Ramanujachary KV, in King RB. (ed.), *Oxides: Solid-State Chemistry in Encyclopedia of Inorganic Chemistry*, 2nd ed. Wiley, pp. 4006–4053, 2005.
2. Sekhar MMA, Combustion synthesis and properties of ferroelectric, relaxor ferroelectrics and microwave resonator materials, PhD Thesis, Indian Institute of Science, 1994.
3. Sekhar MMA, Patil KC, Combustion synthesis and properties of fine-particle dielectric oxide materials, *J Mater Chem* **2**: 739–743, 1992.
4. Anuradha TV, Ranganathan S, Mimani T, Patil KC, Combustion synthesis of nanostructured barium titanate, *Scr Mater* **44**: 2237–2241, 2001.
5. Sekhar MMA, Patil KC, Combustion synthesis of lead-based dielectrics: A comparative study of redox compounds and mixtures, *Int J Self-Propagating High-Temp Synth* **3**: 27–38, 1994.
6. Veierheilig A, Safari A, Halliyal A, Phase stability and dielectric properties of ceramics in the PZN–PNN–PT relaxor ferroelectric system, *Ferroelectrics* **135**: 147–155, 1992.
7. Sekhar MMA, Halliyal A, Patil KC, Combustion synthesis and properties of relaxor materials, *Ferroelectrics* **158**: 289–294, 1994.
8. Sekhar MMA, Halliyal A, Low-temperature synthesis, characterization, and properties of lead-based ferroelectric niobates, *J Am Ceram Soc* **81**: 380–388, 1998.
9. Sekhar MMA, Patil KC, Low temperature synthesis and properties of microwave resonator materials, *Mater Sci Eng* **B 38**: 273–279, 1996.
10. Sundar Manoharan S, Patil KC, Combustion route to fine particle perovskite oxides, *J Solid State Chem* **102**: 267–276, 1993.
11. Patil KC, Fuel cells and SOFC technology, *J Electro Chem Soc of India* **47**: I–XIV, 1998.
12. Gopichandran R, Patil KC, A rapid method to prepare crystalline fine particle chromite powders, *Mater Lett* **12**: 437–441, 1992.
13. Aruna ST, Muthuraman M, Patil KC, Studies on combustion synthesized $LaMnO_3$–$LaCoO_3$, solid solutions, *Mater Res Bull* **35**: 289–296, 2000.
14. Aruna ST, Muthuraman M, Patil KC, Combustion synthesis and properties of strontium substituted lanthanum manganites $La_{1-x}Sr_xMnO_3$ ($0 \leq x \leq 0.3$), *J Mater Chem* **7**(12): 2499–2503, 1997.

15. Suresh K, Panchapagesan TS, Patil KC, Synthesis and properties of $La_{1-x}Sr_xFeO_3$, *Solid State Ionics* **126**: 299–305, 1999.

16. Aruna ST, Muthuraman M, Patil KC, Studies on strontium substituted rare earth manganites, *Solid State Ionics* **120**: 275–280, 1999.

17. Nagabhushana BM, Sreekanth Chakradhar RP, Ramesh KP, Shivakumara C, Chandrappa GT, Low temperature synthesis, structural characterization, and zero-field resistivity of nanocrystalline $La_{1-x}Sr_xMnO_{3+\delta}$ ($0 \le x \le 0.3$) manganites, *Mater Res Bull* **41**: 1735–1746, 2006.

18. Nagabhushana BM, Chandrappa GT, Sreekanth Chakradhar RP, Ramesh KP, Shivakumara C, Synthesis, structural and transport properties of nanocrystalline $La_{1-x}Ba_xMnO_3$ ($0 \le x \le 0.3$) powders, *Solid State Commun* **136**: 427–432, 2005.

19. Nagabhushana BM, Sreekanth Chakradhar RP, Ramesh KP, Shivakumara C, Chandrappa GT, Combustion synthesis, characterization and metal-insulator transition studies of nanocrystalline $La_{1-x}Ca_xMnO_3$ ($0 \le x \le 0.5$), *Mater Chem Phys* **102**: 47–52, 2007.

Chapter 10

Nanocrystalline Oxide Materials for Special Applications

Nanocrystalline oxide materials having selected applications are dealt in this chapter. These include certain simple oxides, rare earth oxides, phosphors, and ceramic pigments. The preparation and properties of calcia, magnesia, yttria, silicates, vanadates, borates, chromites, ceria-based pigments, etc. not covered in the earlier chapters are discussed here.

10.1 SYNTHESIS AND PROPERTIES OF SIMPLE OXIDES

High-purity binary oxides such as CaO and MgO are utilized as specialized materials for refractories. Yttria (Y_2O_3) is another refractory material which possesses a high melting point ($2410°C$) and also does not undergo any phase transformation. Y_2O_3 finds a variety of applications such as high-temperature chemical-resistant substrates, crucibles for melting reactive metals and nozzles for jets, rare-earth-iron magnetic alloys, etc. Insulating oxides such as ZnO are employed in electrical circuits as varistors, a high resistance semiconducting device. They are used to mitigate power surges in electronic circuits and power transmission, and their electrical behavior appears to be determined by grain boundary effects. Y_2O_3 and other rare earth metal oxides are utilized in luminescent materials as a host.

CaO and MgO oxides also draw significant attention as effective chemisorbants for toxic chemicals and gases. Nanocrystalline alkaline earth metal oxides evoke interest as innovative dehalogenation agents. Destructive sorption takes place not only on the surface of these oxide materials, but also in their bulk.

They work in the destructive sorption reaction most efficiently because of their high surface area. In addition, a high concentration of low coordinated sites and structural defects are also found on their surfaces.

Simple binary oxides like CaO, Cr_2O_3, CuO, MgO, NiO, Y_2O_3, and ZnO are prepared by combustion of aqueous solutions containing stoichiometric quantities of the corresponding metal nitrate and fuels like ODH.[1] Yttria is also prepared using other fuels like DFH, MDH, TFTA, and GLY.[2]

Aqueous solutions containing stoichiometric quantities of corresponding metal nitrates and ODH when rapidly heated at 500°C, boil, ignite, and burn to yield voluminous powders. The theoretical equations for the combustion reaction of redox mixtures yielding metal oxides can be written as follows:

$$M(NO_3)_2(aq) + C_2H_6N_4O_2 \text{ (aq)}$$
$$\text{ODH}$$
$$\longrightarrow MO(s) + 2CO_2(g) + 3N_2(g) + 3H_2O(g)$$

(8 mol of gases/mol of MO)

$$M = Cu, Mg, Ni, \text{ and } Zn \tag{1}$$

$$2M(NO_3)_3(aq) + 3C_2H_6N_4O_2 \text{ (aq)}$$
$$\text{ODH}$$
$$\longrightarrow M_2O_3(s) + 6CO_2(g) + 9N_2(g) + 9H_2O(g)$$

(24 mol of gases/mol of M_2O_3)

$$M = Cr \text{ and } Y \tag{2}$$

CaO, MgO, and ZnO are also obtained by the solution combustion reaction of their corresponding metal nitrate and glycine redox mixtures.

$$9M(NO_3)_2(aq) + 10C_2H_5NO_2 \text{ (aq)}$$
$$\text{GLY}$$
$$\longrightarrow 9MO(s) + 20CO_2(g) + 14N_2(g) + 25H_2O(g)$$

($\sim$6.5 mol of gases/mol of MO)

$$M = Ca, Mg, \text{ and } Zn \tag{3}$$

The large amount of gases evolved during combustion not only yield fine-particle oxides but also help to dissipate heat, which inhibits sintering.

Formation of nanocrystalline MgO, ZnO, and Y_2O_3 is confirmed by their characteristic powder XRD patterns (Fig. 10.1).

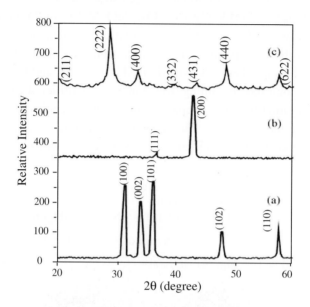

Fig. 10.1.   XRD patterns of as-formed (a) ZnO, (b) MgO, and (c) Y_2O_3.

The composition of the redox mixtures used for combustion and particulate properties of the oxides obtained are summarized in Table 10.1

The TEM micrograph of MgO (Fig. 10.2) shows randomly oriented nanocrystalline particles. The selected-area (SAED) pattern (inset of Fig. 10.2) shows reflection corresponding to (111), (200), (220), and (222) planes indicating the presence of cubic-MgO crystalline phase.

Nanocrystalline MgO having surface area of 107 $m^2\,g^{-1}$ and particle size of 4–11 nm has been investigated for defluoridation of ground water. It is found that nano-MgO powder (0.15 g) has an ability to remove 97% of fluoride from standard sodium fluoride solution (10 ppm) and 75% of fluoride from tube-well water.[3]

Similarly, CaO prepared from glycine with a surface area of 19 $m^2\,g^{-1}$ has been investigated for the removal of chemical oxygen demand (COD) and toxic substances in waste water generated by paper mills.[4] The COD removal capacity of the CaO is found to be up to 93%.

Table 10.1. Compositions of redox mixtures and particulate properties of oxides.

Composition of redox mixtures	Oxide	Powder density $(g\,cm^{-3})$	Surface area $(m^2\,g^{-1})$	Particle size[a] (nm)
$Cu(NO_3)_2 \cdot 6H_2O$ (10 g) + ODH (4.0 g)	CuO	3.78	7	230
$Mg(NO_3)_2 \cdot 6H_2O$ (10 g) + ODH (4.61 g)	MgO	2.86	19	110
$Ni(NO_3)_2 \cdot 6H_2O$ (10 g) + ODH (1.1 g)	NiO	4.93	8	70
$Zn(NO_3)_2 \cdot 6H_2O$ (10 g) + ODH (3.97 g)	ZnO	3.89	35	40
$Cr(NO_3)_3 \cdot 9H_2O$ (8 g) + ODH (3.55 g)	Cr_2O_3	2.86	25	90
$Y(NO_3)_3 \cdot 6H_2O$ (11.5 g) + ODH (5.32 g)	Y_2O_3	4.54	13	102

[a] From surface area.

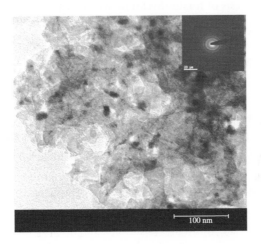

Fig. 10.2. TEM of MgO.

10.2 METAL SILICATES

Forsterite (Mg_2SiO_4), a member of olivine family [$(Mg,Fe)_2SiO_4$] of orthosilicates, along with enstatite $MgSiO_3$ (isosilicate) and other similar minerals constitute the fundamental components of the rocks in the upper mantle of the earth. The recent discovery of tunable lasing property of 0.4 at.%

Cr^{4+}/Cr^{3+}-doped forsterite crystal in the near-infrared range of 1.167–1.345 μm has led to renewed interest in this material.

Calcium silicate (Wollastonite) and magnesium silicate (Forsterite) are prepared by heterogeneous solution combustion method and their properties investigated. Diformyl hydrazide (OHC–HN–NH–CHO) is a potential ligand, and metal complex formation with it increases the solubility of the metal ions; it also prevents their precipitation as water evaporates. DFH is a good fuel like other N–N bonded hydrazine derivatives. In a typical experiment, 15 g $Mg(NO_3)_2 \cdot 6H_2O$, 5.72 g NH_4NO_3, along with 8 g DFH are dissolved in a minimum quantity of water in a cylindrical Petri dish. To this 1.72 g of fumed silica is added and dispersed well. The heterogeneous redox mixture when rapidly heated at 500°C dehydrates, forming a honey-like gel. This ignites by the methathetical reaction between magnesia and homogeneously dispersed silica to yield voluminous forsterite in less than 5 min.[5]

Theoretical equations assuming complete combustion of the redox mixture used for the synthesis of forsterite can be written as

$$4Mg(NO_3)_2(aq) + 2SiO_2(s) + \underset{\text{DFH}}{5C_2H_4N_2O_2\,(aq)}$$

$$\longrightarrow 2Mg_2SiO_4(s) + 9N_2(g) + 10H_2O(g) + 10CO_2(g)$$

$$\text{(14.5 mol of gases/mol of forsterite)} \qquad (4)$$

It is interesting to note that no flame appears during the combustion unless extra amount of redox mixture (NH_4NO_3 + DFH) is added. The presence of dispersed phase (SiO_2) reduces foaming as these oxides become trapped in the pores of the foam. In the presence of NH_4NO_3, the number of moles of gases evolved per mole of forsterite increases and this also alters the redox chemistry of the process.

The powder X-ray diffraction patterns of as-formed and calcined forsterite prepared by combustion and sol–gel processes are shown in Fig. 10.3. The diffraction patterns reflect the single-phase nature of the powders.

Materials with porous architecture and high surface area are being developed for numerous potential areas in nanotechnology. Specific areas of interest include catalysis and separation science. Porous ceramics have a wide range of applications in biomedical engineering, e.g., in the fabrication of cranial plugs in bone reconstruction and the production of synthetic substrates for the fabrication of artificial corneas. There are a variety of such applications and a range of compounds such as wollastonite ($CaSiO_3$) are being made use of. $CaSiO_3$

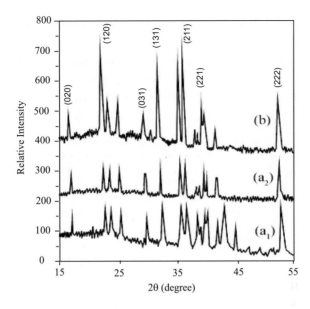

Fig. 10.3. Powder XRD patterns of forsterite (a) combustion process: a_1 — as prepared, a_2 — calcined at 1300°C, (b) sol–gel process: calcined at 1000°C.

has also been used traditionally as a filler of resins or paper, an alternative to asbestos, high-frequency insulators, and as machinable ceramics.

Single-phase β-$CaSiO_3$ and α-$CaSiO_3$ nanocrystalline powders are synthesized by the solution combustion process. $CaSiO_3$ ceramic powder is prepared by dissolving calcium nitrate and diformyl hydrazide ($C_2H_4N_2O_2$) as fuel in minimum quantity of water in a cylindrical Petri dish. Fumed silica is added to this mixture and dispersed well by using a magnetic stirrer for half an hour. The redox mixture when heated at 500°C boils and thermally dehydrates, forming a honeycomb like gel which ignites to yield voluminous $CaSiO_3$ powder.[6] The theoretical equation assuming complete combustion of the redox mixture used for the synthesis of $CaSiO_3$ can be written as

$$4Ca(NO_3)_2(aq) + 4SiO_2(s) + 5C_2H_4N_2O_2 \text{ (aq)}$$
$$\text{DFH}$$
$$\longrightarrow 4CaSiO_3(s) + 9N_2(g) + 10H_2O(g) + 10CO_2(g)$$
$$(\sim7 \text{ mol of gases/mol of } CaSiO_3) \qquad (5)$$

XRD patterns of the as-formed $CaSiO_3$ nanocrystalline powders show peaks corresponding to mixtures of $CaSiO_3$ and CaO. On calcination at 950°C for

3 h, the powders start crystallizing and give rise to a single-phase β-CaSiO$_3$. The peak corresponding to CaO disappears at 950°C. This indicates that the samples undergo complete crystallization with no detectable impurity when calcined at this temperature. The physical properties of the powdered samples of wollastonite (CaSiO$_3$) such as surface area, particle size, shape, and powder density are given in Table 10.2.

Table 10.2.   Particulate properties of wollastonite.

Properties	Results
Powder density	(g cm^{-3})
As formed	1.73
At 950°C/3 h	0.89
At 1250°C/3 h	0.77
Surface area	(m^2 g^{-1})
As formed	31.93
At 950°C/3 h	0.585
At 1250°C/3 h	3.48
Particle size	(nm)
From XRD	29–50
Porosity	(%)
At 950°C/3 h	17.5
At 1250°C/3 h	31.6

The microstructure of the powder is studied by SEM and it is interesting to note that with increase in temperature, the samples become more and more porous and the pore diameter increases from 0.25 to 8 μm. The calculated porosity values are 17.5% and 31.6% at 950°C and 1200°C, respectively. Thermoluminescence (TL) studies show more intensity in the powder when compared to pelletized sample and there is a shift in TL glow peaks as well as reduction in TL intensity of pelletized samples. This is attributed to interparticle spacing and pressure induced defects.

10.3 CERAMIC PIGMENTS

Inorganic metal oxides are used as pigments for paints and coatings, printing inks and coloring agents for plastics, enamels, and ceramics. In 2002 alone, a multibillion dollar ($2.5 billion) market for these materials existed in the

United States. Most pigments used today are synthetic and tailor-made to meet exact standards with respect to color, applications, and weathering. Natural pigments, mainly iron oxides, account for only ~3% of world production. Titania (TiO_2) accounts for two-third of the pigment market. Zinc oxide (ZnO) is used as a colorant and antifungal agent in polymer composites. Other important ceramic oxide pigments are: Cr_2O_3 (green), $CoAl_2O_4$ (cobalt blue), and $(Fe_{1-x}Mn_x)_2O_3$ (black to red brown). A number of rare earth elements are also used as coloring agents in ceramics, e.g., Pr^{4+} in $ZrO_2/ZrSiO_4$ (yellow), $Ce_{1-x}Pr_xO_2$ (red), and Ce_2S_3 (red).

The important physical-optical properties of pigments are their light absorption and scattering properties which depend upon wavelength, particle size, particle shape, and refractive index. Thus, to give good tinctorial strength (ability of a colorant to impart color) colored pigments should have an optimum particle size of 1–10 μm and possess a high refractive index of 1.8–2.0. Apart from this, they should be stable at high temperatures. Pigment particles of nanoscale size are required for special applications such as coloring and UV-stabilization of plastics, inks, transparent thin films on glass, and coatings on luminescent materials. Nanoparticles are able to provide more vivid colors that resist deterioration and fading over time. They also facilitate the uniform dispersion of the pigment into a wide range of material substrates.

The preparation and properties of alumina, zirconia, and zircon-based ceramic pigments have already been discussed in Chaps. 4 and 8. In this section, the preparation and properties of other ceramic pigments are described.

Color chemistry of cobalt. It is interesting to note that colors of cobalt(II) compounds are stereochemically specific depending upon the coordination site and coordination number.[7] While octahedral cobalt compounds are generally pink to violet, the tetrahedral ones are blue. The production of different colors by cobalt compounds and ions is also possible in the solid state. Their colors range from blue, green to pink depending upon the host lattice and coordination geometry (Tables 10.3(a) and 10.3(b)).

10.3.1 *Borate Pigments*

Allochromatic ($Co_xMg_{2-x}B_2O_5$) and idiochromatic ($Co_2B_2O_5$) mauve-pink (purple) pigments prepared by solution combustion synthesis are discussed here. Rosy cobalt pigment is obtained by rapidly heating an aqueous solution

Table 10.3(a).   Various colors of cobalt(II) complexes in solutions.

Coordination compounds of cobalt(II)	Color	Cobalt(II) coordination site
$[Co(H_2O)_6]^{2+}$	Pink	Octahedral
$[Co(NH_3)_5Cl]^{2+}$	Pink	Octahedral
$[CoCl(H_2O)_5]^{2+}$	Raspberry red	Octahedral
$Co(CH_3COO)_2 \cdot H_2O$	Violet red	Octahedral
$[Co(bipy)_3]^{2+}$	Red	Octahedral
CoF_2	Pink	Octahedral
$CoCl_2$	Sky blue	Octahedral
$[CoCl_4]^{2-}$	Blue	Tetrahedral
$[Co(NCS)_4]^{2-}$	Blue	Tetrahedral
$[Co(N_3)_4]^{2-}$	Blue	Tetrahedral
$[(C_2H_5O)_2PS]Co(C_2H_5OH)_2$	Blue	Tetrahedral
$[Co(CN)_5]^{3-}$	Dark green	Tetragonal pyramidal

Table 10.3(b).   Various colors of cobalt(II) in solids oxide materials.

Compounds of cobalt(II)	Color	Cobalt(II) coordination site
Co^{2+} in Al_2O_3	Dark blue	Octahedral
$CoAl_2O_4$	Indigo	Tetrahedral
Co^{2+} in MAl_2O_4	Sky blue	Tetrahedral
Co^{2+} in Zn_2SiO_4	Lavender	Tetrahedral
$Co_2B_2O_5$	Purple	Octahedral
Co^{2+} in $Mg_2B_2O_5$	Rose pink	Octahedral
Co^{2+} in $ZrSiO_4$	Red	Interstial
Co^{2+} in $CoCr_2O_4$	Peacock green	Tetrahedral
Co^{2+} in ZnO	Green	Tetrahedral

containing 20 g $Mg(NO_3)_2 \cdot 6H_2O$, 4.8 g H_3BO_3, and 7.8 g of CH at 350°C. The combustion is flaming type (flame temperature $\sim$1250°C) yielding 17% (2.9 g) of rose pink $Mg_2B_2O_5$. Cobalt pyroborate $Co_2B_2O_5$ can also be prepared by rapidly heating an aqueous solution containing 5 g $Co(NO_3)_2 \cdot 6H_2O$, 1 g H_3BO_3, and 2 g CH at 350°C. This solution undergoes smouldering type combustion (flame temperature $\sim$1000°C) to give a mauve-pink (purple) product.[8]

The theoretical equations assuming complete combustion for the formation of $Co_2B_2O_5$ and $Co_xMg_{2-x}B_2O_5$ can be written as

$$4Co(NO_3)_2(aq) + 4H_3BO_3(aq) + 5CH_6N_4O\,(aq)$$
$$\xrightarrow[\text{CH}]{350°C} 2Co_2B_2O_5(g) + 5CO_2(g) + 14N_2(g) + 21H_2O(g)$$
$$(20 \text{ mol of gases/mol of } Co_2B_2O_5) \quad (6)$$

$$4Mg(NO_3)_2(aq) + 4H_3BO_3(aq) + 5CH_6N_4O\,(aq)$$
$$\xrightarrow[\text{CH} \atop 350°C]{Co(NO_3)_2 \cdot 6H_2O} 2Co_xMg_{2-x}B_2O_5(s) + 5CO_2(g) + 14N_2(g) + 21H_2O(g)$$
$$(20 \text{ mol of gases/mol of } Co_xMg_{2-x}B_2O_5) \quad (7)$$

The color of the allochromatic $Co_xMg_{2-x}B_2O_5$ compares well with the purple color of $Co_2B_2O_5$. The as-synthesized $Co_2B_2O_5$ is X-ray amorphous. The crystalline phase occurs after sintering at 1200°C for 4 h. Powder XRD patterns of $Co_2B_2O_5$ and $Co_xMg_{2-x}B_2O_5$ are similar to triclinic $Mg_2B_2O_5$ (Fig. 10.4). The reflectance spectra of $Co_2B_2O_5$ and $Co_xMg_{2-x}B_2O_5$ are

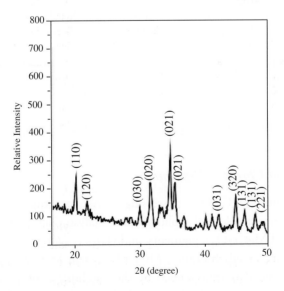

Fig. 10.4.   XRD pattern of as-formed $Co_xMg_{2-x}B_2O_5$.

presented in Fig. 10.5. The spectrum of $Co_xMg_{2-x}B_2O_5$ is similar to that of violet-red $Co(CH_3COO)_2 \cdot 4H_2O$ in which cobalt is octahedrally coordinated by acetate ligands. Octahedral cobalt compounds are generally pink in color. Therefore, in rose pink $Co_xMg_{2-x}B_2O_5$, the cobalt substitutes for magnesium in the octahedral lattice site of the pyroborate structure. The purple $Co_2B_2O_5$ shows a similar but broader electronic spectrum suggesting cobalt in an octahedral environment.

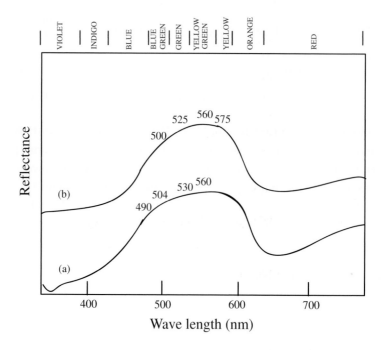

Fig. 10.5.   Diffuse reflectance spectra of (a) $Co_2B_2O_5$ and (b) $Co_xMg_{2-x}B_2O_5$.

10.3.2 *Metal Chromite Pigments*

Metal chromites (MCr_2O_4) possessing spinel structure are of interest because of their technological applications as catalysts ($ZnCr_2O_4$), refractories ($CaCr_2O_4$ and $MgCr_2O_4$), pigments, and glazes ($CoCr_2O_4$).

Fine-particle metal chromites, MCr_2O_4 (where M = Mg, Ca, Mn, Fe, Co, Ni, Cu, and Zn) are obtained by the combustion of a redox mixture containing corresponding divalent metal nitrate, chromium(III) nitrate and tetraformal trisazine (TFTA)[9] and urea.[10] The energetics of the combustion process depends strongly on the choice of the fuel. The choice of the TFTA fuel arose because the energetics of the chromites precludes the use of metal nitrate–urea/carbohydrazide redox mixtures usually employed successfully in making other materials. Consequently, TFTA is the preferred fuel for the preparation of metal chromites.

In a typical experiment, for the preparation of $MnCr_2O_4$, 2 g of manganese nitrate, 6.38 g of chromium nitrate, and 2.05 g of TFTA are dissolved in minimum quantity of water. In the case of urea, the stoichiometric composition of the combustion mixture for the preparation of $CoCr_2O_4$ contains 2.32 g of $Co(NO_3)_2 \cdot 4H_2O$, 6.38 g of $Cr(NO_3)_3 \cdot 9H_2O$, and 3.18 g of urea. The Pyrex dish containing the redox mixture is heated to a temperature of $350 \pm 5°C$. The solution boils and undergoes dehydration followed by ignition and combustion to yield a fine foamy product.

The theoretical equations assuming complete combustion for the formation of MCr_2O_4 can be written as

$$M(NO_3)_2(aq) + 2Cr(NO_3)_3(aq) + \underset{\text{TFTA}}{2C_4H_{16}N_6O_2 \,(aq)} + 4HNO_3(aq)$$

$$\longrightarrow MCr_2O_4(s) + 8CO_2(g) + 18H_2O(g) + 12N_2(g) + O_2(g)$$

$$\text{(39 mol of gases/mol of } MCr_2O_4) \tag{8}$$

$$3M(NO_3)_{2(aq)} + 6Cr(NO_3)_{3(aq)} + \underset{\text{Urea}}{20CH_4N_2O_{(aq)}}$$

$$\longrightarrow 3MCr_2O_{4(s)} + 20CO_{2(g)} + 40H_2O_{(g)} + 32N_{2(g)}$$

$$(\sim31 \text{ mol of gases/mol of } MCr_2O_4)$$

$$M = Mg, Ca, Mn, Fe, Co, Ni, Cu, \text{ and } Zn \tag{9}$$

The actual compositions of the redox mixtures used for the preparation of other metal chromites by TFTA and urea process are given in Table 10.4. The table also summarizes the particulate properties such as green density, particle size, surface area, and lattice constants of the as-prepared chromite powders.

The formation of chromites is confirmed by their characteristic XRD patterns for the as-prepared powders. Typical XRD pattern of MCr_2O_4 is shown in Fig. 10.6. The lattice constants of the chromites calculated from the XRD pattern are in good agreement with the reported values.

Table 10.4.   Compositions of redox mixtures and particulate properties of metal chromites.

Composition of the mixtures	Combustion product (color)		Powder density (g cm^{-3})		Surface area (m^2 g^{-1})		Particle size[a] (nm)	
	TFTA	Urea	TFTA	Urea	TFTA	Urea	TFTA	Urea
2.04 g Mg(NO$_3$)$_2$ · 6H$_2$O + A + B	MgCr$_2$O$_4$ Green	MgCr$_2$O$_4$ Green	3.2	3.25	30.0	15.5	40	100
1.88 g Ca(NO$_3$)$_2$ · 4H$_2$O + A + B	CaCr$_2$O$_4$ Green	CaCr$_2$O$_4$ Green	1.9	2.1	14.5	11.5	—	200
2.0 g Mn(NO$_3$)$_2$ · 4H$_2$O + A + B	MnCr$_2$O$_4$ Green	MnCr$_2$O$_4$ Green	3.7	3.9	18.5	14	30	100
3.22 g Fe(NO$_3$)$_3$ · 9H$_2$O + A + B	FeCr$_2$O$_4$ Brown	FeCr$_2$O$_4$ Green	4.2	4.35	15.0	25.3	90	50
2.32 g Co(NO$_3$)$_2$ · 4H$_2$O + A + B	CoCr$_2$O$_4$ Bluish green	CoCr$_2$O$_4$ Bluish green	4.1	4.1	20.5	8.3	70	150
2.32 g Ni(NO$_3$)$_2$ · 6H$_2$O + A + B	NiCr$_2$O$_4$ Dark green	NiCr$_2$O$_4$ Dark green	3.9	4.0	13.5	9.5	90	150
1.93 g Cu(NO$_3$)$_2$ · 6H$_2$O + A + B/b	CuCr$_2$O$_4$ Black	CuCr$_2$O$_4$ Black	4.5	4.2	20.0	5.75	60	240
2.37 g Zn(NO$_3$)$_2$ · 6H$_2$O + A + B	ZnCr$_2$O$_4$ Green	ZnCr$_2$O$_4$ Green	4.3	4.35	24.5	20.9	60	60

A = 6.38 g Cr(NO$_3$)$_3$ · 9H$_2$O; B = TFTA 2.3 g or Urea 3.18 g; b = 1.54 g TFTA.
[a] From surface area.

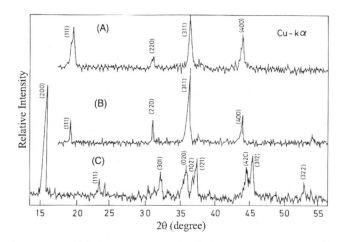

Fig. 10.6. XRD patterns of as-formed (A) $MgCr_2O_4$, (B) $CoCr_2O_4$, and (C) α-$CaCr_2O_4$.

In the case of copper chromite ($CuCr_2O_4$), fuel-lean (oxidizer-rich) mixtures are used to reduce the exothermicity of the combustion reaction and to prevent the decomposition of $CuCr_2O_4$ to $Cu_2Cr_2O_4$ and Cr_2O_3. The as-formed $CuCr_2O_4$ powder however, is X-ray amorphous and when heated at 800°C for 1 h, crystallizes to give the expected XRD pattern. It is interesting to note that the combustion process directly yields α-$CaCr_2O_4$, which is otherwise reported to form only at temperatures above 1600°C. In this case, the formation of such high-temperature phase is attributed to the high *in situ* temperature generated during combustion and to the rapid cooling rate which stabilizes such phases.

10.3.3 *Silicate Pigments*

When metal ions such as Co^{2+} and Ni^{2+} cations replace Zn^{2+} ion isomorphously in tetrahedral sites of the silicate willemite, they impart blue color to the zinc silicate.

In a typical experiment, 29.73 g $Zn(NO_3)_2 \cdot 6H_2O$, 2 mol% Ni/Co $(NO_3)_2$, and 11 g DFH are dissolved in a minimum quantity of water in a cylindrical Petri dish. Three grams fumed silica is added to this mixture and dispersed well. The heterogeneous redox mixture is rapidly heated in a preheated muffle furnace (500 ± 10°C). The solution boils, foams, and ignites to burn with a flame (temperature ~1400°C) yielding voluminous blue colored willemite pigments in less than 5 min.[11] Theoretical equations for the synthesis of willemite assuming complete combustion of the redox mixture can be written as

$$4Zn(NO_3)_2(aq) + 2SiO_2(s) + 5C_2H_4N_2O_2 \text{ (aq)}$$
$$\text{DFH}$$
$$\longrightarrow 2Mn^{2+}/Zn_2SiO_4(s) + 9N_2(g) + 10H_2O(g) + 10CO_2(g)$$
$$(\sim 15 \text{ mol of gases/mol of } Zn_2SiO_4) \qquad (10)$$

Fumed silica controls the violent combustion reaction as the combustible gases get trapped in the pores of the foam.

The phase formation is confirmed by powder XRD pattern. The as-formed powders show willemite phase along with ZnO, which completely disappears on calcination at $\sim$800°C (Fig. 10.7). The ^{29}Si MAS-NMR spectrum of calcined willemite powder shows a single resonance at -66 ppm characteristic of the monomeric orthosilicate SiO_4 unit. The willemite powder has a surface area of 26 m^2 g^{-1}. The reflectance spectrum of $Co^{2+}:Zn_2SiO_4$ reveals a broad intense band in the width range of 460–700 nm, with absorption maxima at

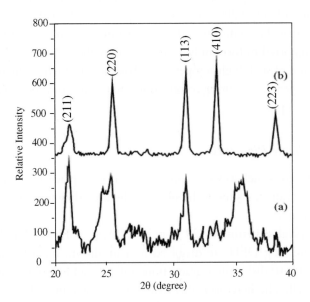

Fig. 10.7.   Powder XRD patterns of $Co^{2+}:Zn_2SiO_4$ (a) as-formed and (b) sintered at 950°C for 1 h.

550, 586, and 640 nm (Fig. 10.8). This could be assigned to $^4A_2(F) \rightarrow {}^4T_1(P)$ d–d transition in the tetrahedral environment. The reflectance spectrum of $Ni^{2+}:Zn_2SiO_4$ exhibits a wide, intense band in the range of 480–760 nm, with absorption maxima at 540, 584, and 633 nm, which corresponds to $^3T_1(F) \rightarrow {}^3T_1(P)$ d–d transition in the tetrahedral environment. In both spectra, the wide band appears to be splitting by Russell–Saunders coupling. The high intensity of these bands is a consequence of the interaction of the "3d" orbitals with the "4p" orbitals of the ligands. In the Zn_2SiO_4 structure, Co^{2+} occupies the two inequivalent Zn^{2+} sites of C_1 symmetry in the willemite structure. The cobalt and nickel ions impart a blue color to the pigments and exhibit three electronic transitions. The first two appear in the IR region and the third band in the visible region.

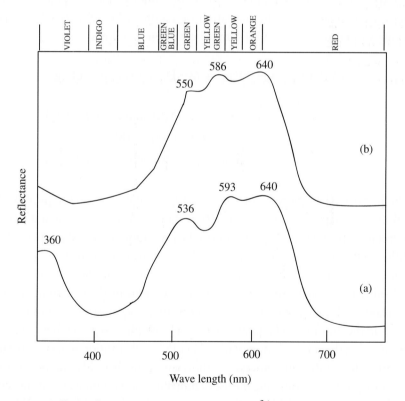

Fig. 10.8. Diffuse reflectance absorption spectra of $Co^{2+}:Zn_2SiO_4$ (a) as-formed and (b) sintered at 950°C for 1 h.

10.3.4   *Ceria-Based Pigment* — $Ce_{1-x}Pr_xO_{2-\delta}$

Cadmium sulfoselenide [Cd(S,Se)] is normally used as red pigment although it is toxic and hazardous to the environment. These pigments are now banned with world-wide steps being taken to replace them by ceria-based red pigment such as cerium sulfide and praseodymium-doped ceria ($Ce_{1-x}Pr_xO_{2-\delta}$). Such substitutions have been successfully tested as inorganic pigments for porcelain stoneware tiles so as to obtain different shades of red and orange hues. Presently, solution combustion synthesis is also used to prepare $Ce_{1-x}Pr_xO_{2-\delta}$ red pigment and its properties examined.[12]

In a typical experiment, $Ce_{0.99}Pr_{0.01}O_2$ pigment is prepared by heating an aqueous solution of 10 g $Ce(NO_3)_3$, 0.0396 g Pr_6O_{11} (dissolved in dil. HNO_3), and 4 g of ODH in a preheated muffle furnace at a temperature of ~350°C. The solution ignites to yield voluminous, fine red pigment. The theoretical equation for the formation of 1 atom% Pr^{4+}:CeO_2 by solution combustion process using ODH may be written as follows:

$$4Ce(NO_3)_2 \cdot 6H_2O(aq) + 6\underset{ODH}{C_2H_6N_4O_2}(aq) + O_2(g)$$

$$\xrightarrow{Pr(NO_3)_3/350°C} 4Ce_{1-x}Pr_xO_{2-\delta}(s) + 12CO_2(g)$$

$$+ 42H_2O(g) + 18N_2(g)$$

$$(18 \text{ mol of gases/mol of } Ce_{1-x}Pr_xO_{2-\delta}) \qquad (11)$$

The powder XRD patterns of $Ce_{1-x}Pr_xO_{2-\delta}$ pigments show typical XRD reflections of cubic ceria phase (Fig. 10.9). The characteristic color and diffuse reflectance spectra (Fig. 10.10) confirm the formation of the red pigment. The diffuse reflectance spectra of pigments show an absorption edge at ~690 nm. The position of the absorption edge depends critically on the Pr content. It is red-shifted with increasing Pr^{4+} content. The slope of the electronic spectra decreases with increasing Pr content, confirming the substitution of Pr^{4+} for Ce^{4+} in the fluorite ceria crystal. Interestingly, no phase separation is observed on sintering the solid solutions with higher Pr^{4+} substitution in CeO_2. In this manner, the combustion process in a short time directly yields a homogeneous nanosize product eliminating the use of mineralizers. This indicates the novelty of the process. The TEM of red pigment particles obtained from ODH fuel is almost spherical and large (Fig. 10.11).

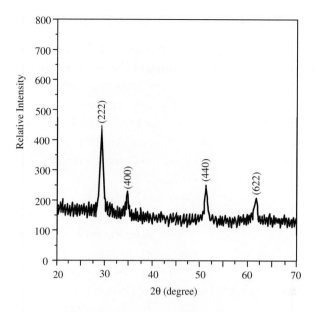

Fig. 10.9. XRD pattern of $Ce_{1-x}Pr_xO_{2-\delta}$ prepared by ODH.

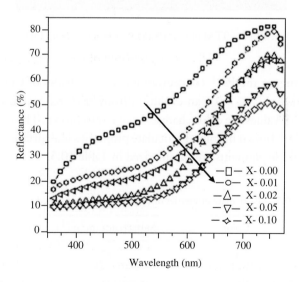

Fig. 10.10. Reflectance spectra of $Ce_{1-x}Pr_xO_{2-\delta}$ with $x = 0\text{--}0.1$.

Fig. 10.11.   TEM of $Ce_{0.9}Pr_{0.1}O_{2-\delta}$ obtained from ODH.

Depending on the Pr^{4+} concentration, various shades of $Ce_{1-x}Pr_xO_{2-\delta}$ are obtained. For $x \leq 0.05$, an evolution from brick red to red is observed. For higher "x" values the color changes from red to brown. The $Ce_{0.5}Pr_{0.5}O_2$ pigment is dark brown in color. Particulate properties like surface area, particle size, etc. of these pigments are summarized in Table 10.5.

Table 10.5.   Particulate properties of $Ce_{1-x}Pr_xO_{2-\delta}$ obtained from ODH.

Pigment	Color	Surface area ($m^2\,g^{-1}$)	Crystallite size (nm)
$Ce_{0.99}Pr_{0.01}O_{2-\delta}$	Brick red	82	31
$Ce_{0.95}Pr_{0.05}O_{2-\delta}$	Brick red	73	35
$Ce_{0.8}Pr_{0.2}O_{2-\delta}$	Red	71	38
$Ce_{0.7}Pr_{0.3}O_{2-\delta}$	Reddish brown	67	40
$Ce_{0.5}Pr_{0.5}O_{2-\delta}$	Brown	—	—

The reflectance spectra of $Ce_{0.95}Pr_{0.05}O_{2-\delta}$ sintered at different temperatures are shown in Fig. 10.12. It is clear from the figure that the absorption edge changes with the sintering temperature and shows red shift. From the reflectance spectra, the values of (L^*, a^*, b^*) are calculated and represented in Figs. 10.13 and 10.14.[13] With the increase in the crystallite size, the pigments become more yellow and red. The darkness of the pigment increases with sintering. It is also seen that sintering at 1000°C for 1 h is optimum for the better color properties of the $Ce_{0.95}Pr_{0.05}O_{2-\delta}$. Praseodymium being a more expensive compound than cerium, doped ceria pigment with a low % of Pr is more economical. Hence, the composition $Ce_{0.98}Pr_{0.02}O_{2-\delta}$ can be tailored for the industrial production of CeO_2-based red ceramic pigment.

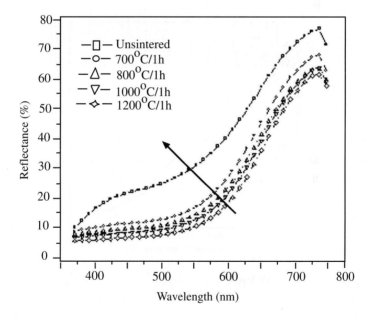

Fig. 10.12.   Reflectance spectra of $Ce_{0.95}Pr_{0.05}O_{2-\delta}$ sintered at different temperature.

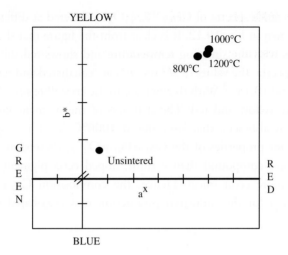

Fig. 10.13.   Effect of Pr doping on the color coordinates (a* and b*) in $Ce_{1-x}Pr_xO_{2-\delta}$.

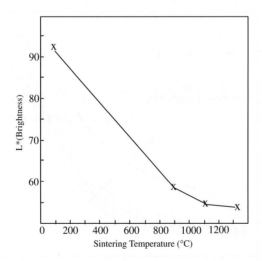

Fig. 10.14.   Effect of Pr doping on the color coordinates (L*) in $Ce_{1-x}Pr_xO_{2-\delta}$.

10.4 Eu^{3+}-ION-DOPED RED PHOSPHORS

Compact fluorescent lamps (CFL) are commonly used these days and contain red, green, and blue emitting phosphors mixed in the ratio of 6:3:1 to emit white light. The preparation of green and blue phosphors was explained earlier in Chap. 4. Here, the preparation and properties of red phosphors is presented.

The activator used for red phosphors is Eu^{3+} (4f^6) and the commonly used hosts are Y$_2$O$_3$, YVO$_4$, and LnBO$_3$ (Ln = La, Gd, and Y).

(i) Eu^{3+}-activated Y$_2$O$_3$ is prepared as follows. An aqueous redox mixture containing stoichiometric amounts of yttrium nitrate, europium nitrate, and ODH is introduced into a muffle furnace preheated to 350–500°C (see Eq. (2)). The mixture undergoes flaming type combustion with the evolution of large amounts of gases.[14] The combustion residue left over is the required red phosphor.

Assuming complete combustion, the theoretical equation for the formation of Eu^{3+}-activated Y$_2$O$_3$ red phosphor using ODH as the fuel may be written as follows:

$$2Y(NO_3)_3(aq) + 3\underset{ODH}{C_2H_6N_4O_2}\,(aq)$$

$$\xrightarrow{Eu^{3+}} Eu^{3+}/Y_2O_3(s) + 9H_2O(g) + 6CO_2(g) + 9N_2$$

$$(24 \text{ mol of gases/mol of } Y_2O_3) \quad (12)$$

Similarly, red phosphor is also prepared using urea as fuel. The nature of combustion in this case is smoldering type.

(ii) In case of yttrium vanadate red phosphor preparation, ammonium meta vanadate is used as source for vanadium. Consequently, ammonium nitrate is used as oxidizer and 3-methyl-pyrazole-5-one (3MP5O) is used as fuel.[15]

$$4Y(NO_3)_3(aq) + 4NH_4VO_3(aq) + 6NH_4NO_3(aq) + 3\underset{3MP5O}{C_4H_6N_2O}\,(aq)$$

$$\xrightarrow{Eu^{3+}} 4Eu^{3+}/YVO_4(s) + 17N_2(g) + 12CO_2(g) + 29H_2O(g)$$

$$(\sim15 \text{ mol of gases/mol of } YVO_4) \quad (13)$$

(iii) Eu^{3+}-activated borate is prepared by combustion reaction of an aqueous redox mixture containing stoichiometric amounts of Ln(NO$_3$)$_3$, H$_3$BO$_3$, and

ODH.[14] Since boric acid is a neutral compound, no extra fuel or oxidizer is required. Assuming complete combustion, the formation of borate may be represented by the following reactions:

$$2Ln(NO_3)_3(aq) + 3C_2H_6N_4O_2(aq)$$
$$\text{ODH}$$
$$\longrightarrow Ln_2O_3(s) + 9H_2O(g) + 6CO_2(g) + 9N_2(g)$$
$$(24 \text{ mol of gases/mol of } Ln_2O_3)$$
$$(Ln = La, Gd, and Y) \qquad\qquad (14)$$

$$4Ln(NO_3)_3(aq) + 3C_4H_6N_2O(aq)$$
$$\text{3MP5O}$$
$$\longrightarrow 2Ln_2O_3(s) + 9H_2O(g) + 12CO_2(g) + 9N_2(g)$$
$$(15 \text{ mol of gases/mol of } Ln_2O_3) \qquad (15)$$

$$2H_3BO_3(aq) \longrightarrow B_2O_3(s) + 3H_2O(g) \qquad\qquad (16)$$

The *in situ* formed Ln_2O_3 and B_2O_3 further react to give corresponding metal borates.

$$Ln_2O_3 + B_2O_3(s) \longrightarrow 2LnBO_3(s) \qquad\qquad (17)$$

The powder XRD patterns of $Y_{2-x}Eu_xO_3$ prepared by both ODH and urea processes are shown in Fig. 10.15. The XRD pattern of the phosphor prepared using ODH shows very sharp peaks, as is obvious from the flaming type of combustion observed with ODH fuel. The XRD pattern of the phosphor with urea shows a broader XRD pattern. It is evident that by changing the nature of the fuel one can control the particulate properties. The surface area of the red phosphor prepared by ODH process is $13 \, m^2 \, g^{-1}$, and for the powder prepared by urea process is $22 \, m^2 \, g^{-1}$.

The powder XRD patterns of combustion prepared $LnBO_3$ (ODH process) are shown in Fig. 10.16. The XRD patterns of as-prepared $LnBO_3$ reveals that, except for $GdBO_3$, all other borates are single phase. $GdBO_3$ as a single phase is obtained by calcining it at 900°C for 1 h. As formed YVO_4 is X-ray amorphous, and on heating at 900°C for 1 h shows crystalline phase (Fig. 10.17).

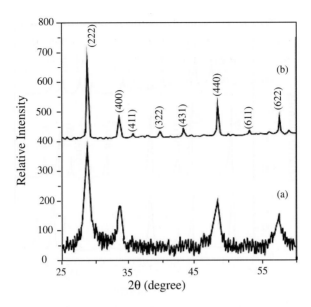

Fig. 10.15.   XRD patterns of Y_2O_3 prepared by (a) urea and (b) ODH.

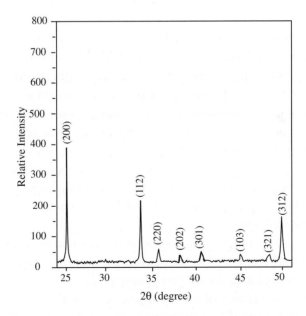

Fig. 10.16.   XRD pattern of Eu^{3+}:YVO_4 red phosphor (3MP5O process) heat treated at 650°C.

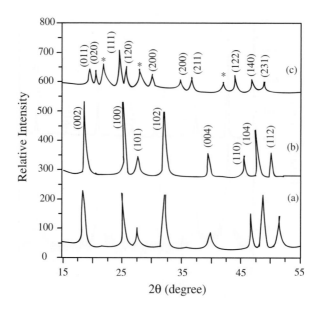

Fig. 10.17. XRD patterns of combustion prepared $LnBO_3$ (ODH process): (a) Ln = Y, (b) Ln = La, and (c) Ln = Gd, where* monoclinic $GdBO_3$.

The emission spectrum of $Y_{1.95}Eu_{0.05}O_3$ (ODH) is shown in Fig. 10.18. The characteristic red-orange emission at 611 nm confirms that Eu^{3+} occupies the C_2 crystallographic site. The emission at 611 nm is caused by electric dipole transition. It is very sharp, looks like atomic emission and is attributed to the electric dipole transition $^5D_0 \rightarrow {}^7F_2$ of the Eu^{3+} ions. The fluorescence spectra of Eu^{3+}-doped $LnBO_3$ shows characteristic red orange emission revealing two bands at 615 and 595 nm. These bands are attributed to $^5D_0 \rightarrow {}^7F_2$ and $^5D_0 \rightarrow {}^7F_1$ transition of nine-coordinated Eu^{3+} ions, respectively (Fig. 10.19). Fluorescence spectrum of $Y_{0.95}Eu_{0.05}VO_4$ is shown in Fig. 10.20. The bands at 611, 615, and 619 nm result from $^5D_0 \geq {}^7F_2$ transition for Eu^{3+}ion.

The effect of calcination temperature on the emission intensity was examined. It is found that with increasing calcination temperature, the emission intensity increases for all the red phosphors.

The effect of addition of Ca^{2+}, Mg^{2+}, Zn^{2+}, Ce^{4+}, Zr^{4+}, and Gd^{3+} (2.5 mol%) in Y_2O_3:Eu^{3+} on the luminescence of the red phosphor was also

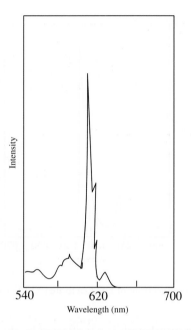

Fig. 10.18.   Emission spectrum of $Y_{1.95}Eu_{0.05}O_3$ (ODH process) at excitation wave length of 254 nm.

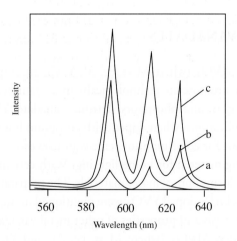

Fig. 10.19.   Fluorescence spectra of Eu^{3+}-activated $LnBO_3$ red phosphor. (a) Ln = La, (b) Ln = Gd, and (c) Ln = Y at excitation wave length of 254 nm.

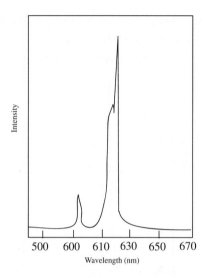

Fig. 10.20.   Fluorescence spectrum of Eu^{3+}-activated YVO_4 red phosphor (3MP5O).

studied. The emission wavelength is not affected by the addition of impurity ions. It is observed that addition of divalent metal ions enhances the emission intensity and tetravalent metal ions decrease the emission intensity.

10.5 METAL VANADATES

Metal vanadates MVO_4 (where $M \equiv Fe$, Al, Y, Cr, etc.), having triclinic or tetragonal crystal structure are technologically important because of their use as laser hosts, masers, and phosphors. Yttrium vanadate, YVO_4, has several technological applications; when doped with europium it forms a highly efficient red phosphor for color monitor screens in television and computers; it also is a promising material for home lighting. With certain rare earth oxide additions, YVO_4 forms optical maser materials. The preparation and luminescent properties of Eu^{3+}-doped YVO_4 were mentioned in the previous section. Presently, the synthesis and properties of other metal vanadates are given.

Metal vanadates, MVO_4 (where $M = Fe$, Al, and Y), are prepared by rapidly heating aqueous solutions containing stoichiometric amounts of metal nitrates (Fe, Al, and Y), ammonium metavanadate, ammonium nitrate, and 3-methylpyrazole-5-one (3MP5O) at 370°C.[15] The redox mixture undergoes a self-propagating, gas producing, exothermic reaction to yield voluminous powder containing fine particles of metal vanadates. The formation of metal vanadates by solution combustion process may be represented by the following reaction sequence.

$$4M(NO_3)_3(aq) + 4NH_4VO_3(aq) + 6NH_4NO_3(aq) + \underset{3MP5O}{3C_4H_6N_2O}\,(aq)$$

$$\longrightarrow 4MVO_4(s) + 17N_2(g) + 12CO_2(g) + 29H_2O(g)$$

$$(\sim 15 \text{ mol of gases/mol of } MVO_4)$$

$$M = Al, Fe \text{ and } Y \tag{18}$$

The large amount of gases evolved during combustion yield foamy, voluminous, and fine-particle metal vanadates. The as-formed combustion products are X-ray amorphous and on calcination at 650°C for 1 h show crystalline single-phase material (Fig. 10.21). The particulate properties are summarized in Table 10.6.

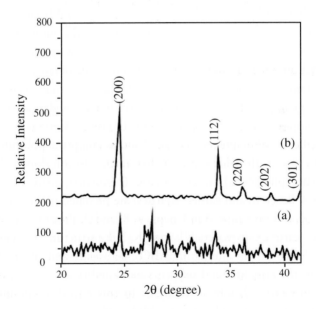

Fig. 10.21.   XRD patterns of (a) FeVO$_4$ and (b) YVO$_4$ heat treated at 650°C for 1 h.

Table 10.6.   Particulate properties of metal vanadate.

Metal vanadate	Powder density ($g\,cm^{-3}$)	Surface area ($m^2\,g^{-1}$)	Particle size[a] (μm)
$FeVO_4$	2.370	2.0	0.912
$AlVO_4$	2.561	7.0	0.335
YVO_4	2.979	14.0	0.144

[a] From surface area.

The main advantage of using solution combustion synthesis process for producing metal vanadates is the saving in energy and decrease in processing time. The fuel, 3MP5O, appears to be ideal for this combustion synthesis.

10.6 RARE EARTH METAL OXIDES (La_2MO_4)

Rare earth metal oxides are technologically important materials as a result of their optical and electronic properties. They are also utilized as hosts for luminescent materials. Certain rare earth oxides find application as core materials for carbon arcs to increase luminosity and spectral balance. A number of them are used as coloring agents in ceramics, the more notable being praseodymium oxide which imparts a brilliant yellow to zirconium silicate glazes. Small amounts are also used in the manufacture of rare earth iron garnet magnets. A mixture of heavier rare earth oxides such as cerium neodymium and didymium oxides are utilized to reduce discoloration in glass screens of color TV and computer monitors caused by iron(III) oxide. Ceria is also employed as an opacifier in white porcelain enamel and as an optical polishing agent.[16]

Various rare earth oxide materials like rare earth ion doped aluminates, borates, yttria (phosphors), zirconia, zircons (pigments), perovskites ($LnMO_3$, M = Al, Cr, Fe, etc.), and garnets (ferrites) are prepared by solution combustion method and investigated in Chaps. 4, 8, and 9. Presently, an interesting class of rare earth oxide materials having K_2NiF_4 structure is discussed. Such mixed oxides consist of alternating ABO_3 perovskite and AO rock salt layers wherein the structure and oxygen stoichiometry can be tuned to study structure–property relations in catalysis. In this regard, catalytic oxidation of ammonia to nitric oxide over La_2MO_4 (M = Co, Ni, Cu) is investigated. Moreover, metallic behavior can be induced in these compounds by doping Ca, Sr, or Ce forming $La_{1.8}Ca_{0.2}CuO_4$, $La_{1.8}Sr_{0.2}CuO_4$ (p-type), and $Nd_{1.85}Ce_{0.15}CuO_4$ (n-type) compounds, respectively.

Since the discovery of superconductivity in multicomponent ceramic oxides like $Ln_{2-x}M_xCuO_4$ (2-1-4), $YBa_2Cu_3O_{7-x}$ (1-2-3), $Bi_2CaSr_2CuO_8$ (2-1-2-2), etc. there has been a tremendous spurt of activity in the preparative techniques of such oxide materials. These compounds are characterized by a layered structure and contain two-dimensional (2D) sheets of Cu–O pyramids or octahedrons in which the charge carriers are holes or electrons. Three distinct but closely related structures (T-, T'-, and T-) have been reported in the (214) system depending on the Ln or dopant content. While T- and T-phases possess apical oxygen atoms that are associated with CuO_6 octahedra and CuO_5 pyramids, respectively, the T'- phase consists of square planar CuO_4 arrangement without apical oxygen atom. Strontium-substituted cuprates show superconductivity.

Fine-particle oxides such as La_2MO_4 are prepared by solution combustion of the corresponding metal nitrates–tetraformal trisazine (TFTA) mixtures. In a typical experiment, 20.0 g lanthanum nitrate, 5.58 g copper nitrate, and 6.95 g TFTA are dissolved in minimum quantity of distilled water in a cylindrical Pyrex dish of 300 ml capacity. The redox mixture when heated on a hot plate maintained at 350°C rapidly evaporates undergoing dehydration. This is followed by foaming and ignition to yield a voluminous oxide. The solid combustion product on annealing at 550°C for 30 min gives crystalline La_2MO_4.[17] The theoretical equation assuming complete combustion for this reaction can be written as follows:

$$2La(NO_3)_3(aq) + M(NO_3)_2(aq) + \underset{\text{TFTA}}{2C_4H_{16}N_6O_2(aq)} + 4HNO_3(aq)$$

$$\longrightarrow La_2MO_4(s) + 8CO_2(g) + 18H_2O(g) + 12N_2(g) + O_2(g)$$

$$(39 \text{ mol of gases/mol of } La_2MO_4)$$

$$(M = Co, Ni, \text{ and } Cu) \tag{19}$$

Rare earth cuprates are also similarly obtained by rapidly heating a saturated aqueous solution containing stoichiometric amounts of corresponding metal nitrates (oxidizer) and MH fuel ($CH_4N_2O_2$,) in a muffle furnace at 350°C. For di- and trivalent metal nitrates, the fuel is taken in the mole ratio of 1:0.3 and 1:0.4, respectively.[18] Metal nitrates are prepared by dissolving the oxides (Sr and Ca are taken as carbonates) in dilute nitric acid. The excess acid is removed by evaporating the solution on a water bath. Theoretical equations assuming complete combustion for the formation of cuprates can be written as follows:

$$2Ln(NO_3)_3(aq) + Cu(NO_3)_2(aq) + 3C_4H_4N_2O_2 (aq)$$
$$\text{MH}$$

$$\xrightarrow{2O_2} Ln_2CuO_4(s) + 12CO_2(g) + 6H_2O(g) + 7N_2(g)$$

$$(25 \text{ moles of gases/mole of } Ln_2CuO_4)$$

$$(Ln = La \text{ and } Nd) \tag{20}$$

Strontium-substituted cuprate ($La_{1.8}Sr_{0.2}CuO_4$) is similarly prepared by the combustion of a redox mixture containing 16.13 g of lanthanum nitrate, 1.17 g of strontium nitrate, 5.0 g of copper nitrate, and 5.19 g of TFTA. The combustion of redox mixture for the preparation of La_2MO_4 and Ln_2CuO_4 results in a voluminous and foamy combustion residue. The powder so obtained is found to be X-ray amorphous.[19] When calcined at 700°C for 30 min, it gives a crystalline single-phase product (Fig. 10.22).

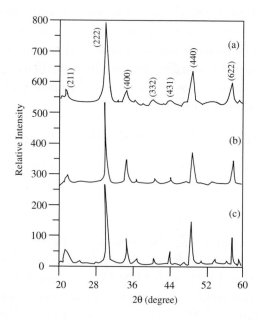

Fig. 10.22.   XRD patterns of heat treated (a) La_2CoO_4, (b) La_2NiO_4, and (c) La_2CuO_4.

La$_2$MO$_4$ oxides are investigated for their catalytic oxidation by making a test run on the N$_2$O decomposition reaction. La$_2$NiO$_4$ shows the highest catalytic activity of 60% for N$_2$O decomposition at 380°C with a rate constant of 1.5×10^{-2} s^{-1} and activation energy of 6.6 kcal mol^{-1}. These results are further utilized in the investigation of catalytic oxidation of ammonia to nitric oxide. The temperature-programmed desorption (TPD) studies made on these oxides show typical TPSR profiles for ammonia oxidation (Fig. 10.23). In all the three cases, nitric oxide and water are the exclusive products formed during oxidation of ammonia to nitric oxide. The onset temperature for NO formation over these oxides are about 275°C, 350°C, and 420°C, for M = Co, Ni, and Cu, respectively.

The BET surface areas of La$_2$CuO$_4$ and La$_{1.8}$Sr$_{0.2}$CuO$_4$ are found to be 1.7 and 3.6 m^2 g^{-1}, respectively. The particle size measured by TEM (Fig. 10.24) shows particles to be below 300 nm. The SEM (Fig. 10.25) of La$_2$CuO$_4$ sintered at 1050°C for 2 h reveals fine grain growth with an average grain size of 5 μm. The density of the sintered pellet is found to be above 95% of the theoretical value.

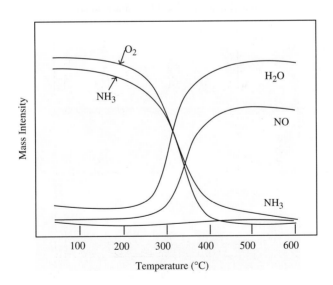

Fig. 10.23.   TPSR profile of catalytic oxidation of ammonia over La$_2$CoO$_4$.

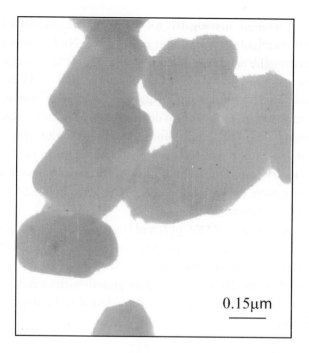

Fig. 10.24.   TEM of $La_{1.8}Sr_{0.2}CuO_4$.

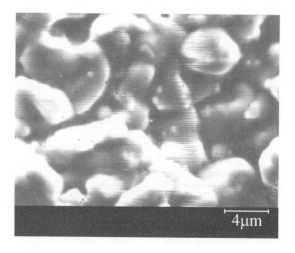

Fig. 10.25.   SEM micrograph of $La_{1.8}Sr_{0.2}CuO_4$ sintered in air at 1050°C for 2 h.

Figure 10.26 illustrates resistivity of $La_{1.8}Sr_{0.2}CuO_4$ as a function of temperature. The room temperature resistivity is 1 mΩ cm. The resistivity curve shows metallic behavior from 300 K to below 50 K and then a drop in resistance to the superconducting state. The onset of the resistance drop is 40 K, the midpoint of the transition is 36 K, while zero resistance is observed at 33 K. The 10–90% transition width is 5.5 K. Figure 10.26 shows the broadening of the transition in various magnetic fields. On the other hand, the low-temperature side depends strongly on the magnetic field strength and the transition is extended towards lower temperatures. The zero resistance of the sample is achieved at 16 K for 2 T and 13 K for 5 T.

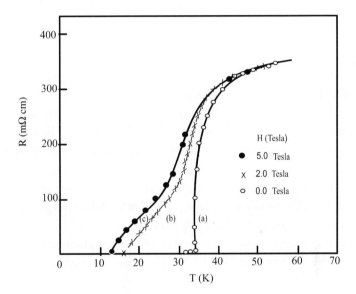

Fig. 10.26. Plot of resistivity versus temperature at different magnetic fields for $La_{1.8}Sr_{0.2}CuO_4$.

The lattice constants calculated for the heat-treated cuprates are given in Table 10.7.

It can be seen that the CuO_4 square of the T'-structure ($Nd_{1.85}Ce_{0.15}CuO_4$) expands considerably, while the "c" axis shrinks in comparison to the T-phase ($La_{1.85}Sr_{0.15}CuO_4$) structure.

When solution combustion approach is used for the preparation of double layer cuprate like $La_{1.6}Sr_{0.4}CaCu_2O_6$ the presence of small amounts

Table 10.7.   Lattice parameters of heat-treated cuprates.

Cuprates	Lattice parameters (nm)
La_2CuO_4	$a = 0.5353$ $b = 0.5392$ $c = 1.3462$
$La_{1.85}Sr_{0.15}CuO_4$	$a = b = 0.3774$ $c = 1.3243$
$La_{1.8}Ca_{0.2}CuO_4$	$a = b = 0.3786$ $c = 1.3156$
Nd_2CuO_4	$a = b = 0.3939$ $c = 1.2182$
$Nd_{1.85}Ce_{0.15}CuO_4$	$a = b = 0.3946$ $c = 1.2044$

of impurity (214) phase is seen. However, this method gives single-phase Sm_2CuO_4, Pr_2CuO_4, and Ce-doped compound.

The microstructure of the sintered La_2CuO_4 (1050°C, 2 h) samples prepared by combustion method has grain size of 4–5 μm. The Ca- and Sr-doped samples have smaller grain size (2–3 μm) than the parent compound (Fig. 10.27). Similar type of behavior is observed in Nd_2CuO_4-and Ce-doped sample. The fully dense (>96% theoretical density) nature of the sintered sample reveals that narrow particle size distribution facilitates greater reactivity of the combustion-derived starting material.

Temperature dependence of resistance of the oxygen-annealed $La_{1.8}Sr_{0.2}CuO_4$ sample is shown in Fig. 10.28. The room temperature resistivity is

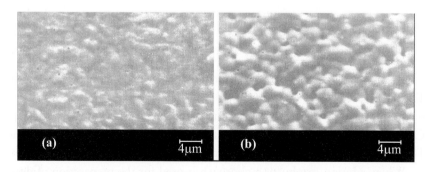

Fig. 10.27.   SEM of sintered (a) $La_{1.8}Ca_{0.2}CuO_4$ and (b) $La_{1.8}Sr_{0.2}CuO_4$.

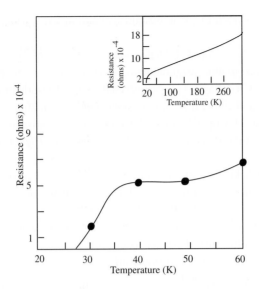

Fig. 10.28.   Temperature dependence of resistance of the oxygen annealed $La_{1.8}Sr_{0.2}CuO_4$.

$1600\ \mu\Omega$ cm. The sample shows metallic behavior in the temperature range of 295–36 K (inset). At 36 K, a sharp superconducting (onset) transition is seen and zero resistance is observed at 28 K.

10.7 CONCLUDING REMARKS

A variety of nanocrystalline simple oxides and doped rare earth oxides prepared by solution combustion method have been investigated as nanopigments, nanophosphors, and superconducting materials. Nanosize metal silicates prepared by this route are well suited for porous ceramics applications. Nano-MgO and CaO have been explored for defluoridation of ground water and treatment of paper mill effluent.

References

1. Suresh K, Patil KC, A recipe for an instant synthesis of fine particle oxide materials, in Rao KJ (ed.), *Perspectives in Solid State Chemistry*, Narosa Publishing House, New Delhi, pp. 376–388, 1995.

2. Ekambaram S, Patil KC, Combustion synthesis of yttria, *J Mater Chem* **5**: 905–908, 1995.

3. Nagappa B, Chandrappa GT, Defluoridation of bore well water using combustion derived nanocrystalline magnesium oxide, *Trans Ind Ceram Soc* **64**: 87–93, 2005.

4. Nagappa B, Chandrappa GT, Nanocrystalline CaO adsorbent to remove COD from paper mill effluent, *J Nanosci Nanotechnol* **7**: 1–4, 2007.

5. Chandrappa GT, Gopi Chandran R, Patil KC, Comparative study of combustion and sol-gel synthesis of forsterite. *Int J Self-Propagating High-Temp Synth* **4**: 183–191, 1995.

6. Sreekanth Chakradhar RP, Nagabhushana BM, Chandrappa GT, Ramesh KP, Rao JL, Solution combustion derived nanocrystalline macroporous wollastonite ceramics. *Mater Chem Phys* **95**: 169–175, 2006.

7. Samrat G, Fine ceramic pigments combustion synthesis and properties, PhD Thesis, Indian Institute of Science, Bangalore, 1998.

8. Mimani T, Samrat G, Combustion synthesis of cobalt pigments: Blue and pink, *Curr Sci* **7**: 892–896, 2000.

9. Sundar Manoharan S, Kumar NRS, Patil KC, Preparation of fine particle chromites: A combustion approach, *Mater Res Bull* **25**: 731–738, 1990.

10. Sundar Manoharan S, Patil KC, Combustion synthesis of metal chromite powders, *J Am Ceram Soc* **75**: 1012–1015, 1992.

11. Chandrappa GT, Ghosh S, Patil KC, Synthesis and properties of willemite, Zn_2SiO_4 and $M^{2+}:Zn_2SiO_4$ (M = Co and Ni), *J Mater Synth Process* **7**: 273–279, 1999.

12. Aruna ST, Ghosh S, Patil KC, Combustion synthesis and properties of $Ce_{1-x}Pr_xO_{2-\delta}$ red ceramic pigments, *Int J Inorg Mater* **3**: 387–392, 2001.

13. Angadi B, Patil KC, Umarji AM, Effect of Pr doping and sintering on the color evolution of combustion prepared $Ce_{1-x}Pr_xO_2$ red pigment, 15th *Annual General Meeting of Materials Research Society of India (MRSI)*, February 9–11, Banaras Hindu University, Varanasi, 2004.

14. Ekambaram S, Patil KC, Maaza M, Synthesis of lamp phosphors: Facile combustion approach, *J Alloys Compd* **393**: 81–92, 2005.

15. Ekambaram S, Patil KC, Rapid synthesis and properties of $FeVO_4$, $AlVO_4$, YVO_4 and Eu^{2+} doped YVO_4, *J Alloys Compd* **217**: 104–107, 1995.

16. Mc Carrol WH, Ramanujachary KV, Oxides: Solid-state chemistry, in King RB (ed.), *Encyclopedia of Inorganic Chemistry* (2nd ed.), Wiley, pp. 4006–4053, 2005.

17. Sundar Manoharan S, Prasad V, Subramanyam SV, Patil KC, Combustion synthesis and properties of fine particle La_2CuO_4 and $La_{1.8}S_{0.2}CuO_4$, *Physica C* **190**: 225–228, 1992.

18. Gopi Chandran R, Patil KC, Combustion synthesis of rare earth cuprates, *Mater Res Bull* **27**: 147–154, 1992.

19. Ramesh S, Manoharan SS, Hegde MS, Patil KC, Catalytic oxidation of ammonia to nitric oxide over La_2MO_4 (M = Co, Ni, Cu) oxides, *J Catal* **157**: 749–751, 1995.

Appendix A

The oxidizers and fuels used in combustion synthesis of oxide materials have been described here.

A.1 OXIDIZERS (METAL NITRATES)

Readily available metal nitrates, ammonium nitrate and ammonium perchlorate are used as such. Wherever metal nitrates are not readily available or are expensive, known amounts of the corresponding metal oxides or carbonates are dissolved in a minimum quantity of dilute nitric acid and the solution is used for combustion.

As titanyl nitrate exists only in solution and is not readily available; its preparation is described below.

A.1.1 *Preparation of Titanyl Nitrate ($TiO(NO_3)_2$)*

Titanyl nitrate is prepared by dissolving freshly precipitated titanyl hydroxide, $TiO(OH)_2$ in dilute nitric acid. Titanyl hydroxide is obtained by the hydrolysis of either titanium tetrachloride ($TiCl_4$) or titanyl iso-propoxide, $Ti(i\text{-}OPr)_4$.[1,2]

$$Ti(i\text{-}OC_3H_7)_4 + 3H_2O \longrightarrow TiO(OH)_2 + 4C_3H_7OH \qquad (A.1)$$

$$TiCl_4 + 2NH_4OH + H_2O \longrightarrow TiO(OH)_2 + 2NH_4Cl + 2HCl \qquad (A.2)$$

$$TiO(OH)_2 + 2HNO_3 \longrightarrow TiO(NO_3)_2 + 2H_2O \qquad (A.3)$$

Caution.   Titanyl tetrachloride is a fuming liquid and hydrolyses in the presence of moisture forming HCl gas and $TiOCl_2$. Therefore titanyl hydroxide is prepared under ice-cold condition ($4°C$).

A.2 FUELS

Readily available fuels like urea, glycine, metal acetates, and ammonium acetate were used as such. The preparation of hydrazine based fuels is described in following sections.

A.2.1 *Carbohydrazide (CH), CH_6N_4O*

Carbohydrazide (CH) is prepared by the reaction of 1 mol of diethyl carbonate with 2 mol of hydrazine hydrate. The reaction is as follows:

$$H_5C_2O–CO–OC_2H_5 + 2N_2H_4 \cdot H_2O$$
$$\longrightarrow H_3N_2–CO–N_2H_3 + 2C_2H_5OH + 2H_2O \qquad (A.4)$$

Preparation.   Three hundred and fifty four grams of diethyl carbonate (3 mol) and 388 g of 99% $N_2H_4 \cdot H_2O$ (6 mol) is placed in a 1 L round bottom flask, fitted with a thermometer. The reactants are partially miscible initially but the flask is shaken well until a single phase is formed. This is accompanied by the evolution of heat causing the temperature to rise to about $55°C$. The flask is then connected through a standard taper joint to a fractionating column filled with raschig rings. A still head fitted with a thermometer and a water cooled condenser is attached to the fractionating column. A heater regulated by a variable transformer is employed to heat the reaction mixture.

    Distillation of ethanol and water (by products) is quite rapid (5 ml/min) for the first 30 min and decreases as the reaction proceeds. Heating is continued for 4 h during which around 325–350 ml of distillate should collect at a vapor temperature of $80–85°C$. The pot temperature is raised from $96°C$ to $119°C$ and then the liquor is cooled to $20°C$ and allowed to stand for at least 1 h. The formed crystals of carbohydrazide are separated by filtration and dried (yield: 165 g, 60%, m.p. $153°C$).[3]

A.2.2 *Oxalyl Dihydrazide (ODH), $C_2H_6N_4O_2$*

Oxalyl dihydrazide (ODH) is prepared by the reaction of 1 mol of diethyl oxalate with 2 mol of hydrazine hydrate.

$$H_5C_2O-CO-CO-OC_2H_5 + 2N_2H_4 \cdot H_2O$$
$$\longrightarrow H_3N_2-CO-CO-N_2H_3 + 2C_2H_5OH + 2H_2O \quad (A.5)$$

Preparation. 146.14 g of diethyl oxalate (1 mol) is added drop wise to 100.12 g of hydrazine hydrate (2 mol) and dissolved in 225 ml of double distilled water in a 1000-ml beaker. The entire addition is carried out in an ice cold bath with vigorous stirring. A white precipitate is obtained and allowed to stand overnight. It is then washed with ethanol, filtered and dried (yield: 100.3 g, 85%, m.p. 239–243°C).[4]

A.2.3 *Tetraformal Trisazine (TFTA), $C_4H_{16}N_6O_2$*

Tetraformal Trisazine (TFTA) is prepared by the reaction of 4 mol of formaldehyde with 3 mol of hydrazine hydrate.

$$4HCHO + 3N_2H_4 \cdot H_2O \longrightarrow C_4H_{16}N_6O_2 + 5H_2O \quad (A.6)$$

Preparation. One hundred milliliters of hydrazine hydrate ($N_2H_4 \cdot H_2O$ 99–100%) is cooled in an ice-salt mixture (temperature of the solvent being 0–5°C). To this ice cold hydrazine hydrate, 175 ml of formaldehyde (HCHO) (37–41%) is added drop wise from a burette. Throughout the addition, the temperature is maintained below 5°C. After the complete addition of HCHO, the mixture is concentrated on a water bath. When white precipitate begins to appear, the solution is removed from the water bath and kept for crystallization. White lumps of solid appear which are filtered, washed with ethanol, and stored in a desiccator. The mother liquor is again concentrated on a water bath for further crop (yield: 70%, m.p. 85°C).[5]

A.2.4 *N, N′-Diformyl Hydrazine (DFH), $C_2H_4N_2O_2$*

Diformyl hydrazine (DFH) is prepared by the reaction of 2 mol of formic acid with 1 mol of hydrazine hydrate.

$$2HCOOH + N_2H_4 \cdot H_2O \longrightarrow OHC-HN-NH-CHO + 3H_2O \quad (A.7)$$

Preparation.   Fifty milliliters of hydrazine hydrate (1 mol) and 150 ml formic acid (2 mol) are mixed and heated overnight on a steam bath and the solvent removed by heating under reduced pressure. Ethanol (100 ml) is added and the solid separated is collected and air dried (yield: 60%, m.p. 160°C).[6]

A.2.5 *Maleic Hydrazide (MH), $C_4H_4N_2O_2$*

Maleic hydrazide (MH) is prepared by the reaction of 1 mol of maleic anhydride with 1 mol of hydrazine chloride.

$$C_4H_2O_3 + N_2H_6Cl_2 \longrightarrow C_4H_4N_2O_2 + HCl + 2H_2O \qquad (A.8)$$

Preparation.   68.6 g of maleic anhydride (0.7 mol) is added to a boiling solution of 75.6 g of $N_2H_6Cl_2$ (0.7 mol) in 500 ml of water, with stirring. The temperature of the mixture is maintained at its boiling point for 3 h. As the volume of the solution is reducing during this period, water is to be added from time to time to prevent evaporation to dryness. After dilution to about 500 ml, crystallization is allowed to proceed to room temperature. A white solid is obtained (yield: 66.8 g, 85% and m.p. 300°C).[7]

A.2.6 *Malonic Acid Dihydrazide (MDH), $C_3H_8N_4O_2$*

Malonic acid dihydrazide (MDH) is prepared by the chemical reaction of 1 mol of diethyl malonate with 2 mol of hydrazine hydrate. The chemical reaction is written as follows:

$$H_2C_2O-CO-CH_2-CO-OC_2H_5 + 2N_2H_4 \cdot H_2O$$
$$\longrightarrow H_3N_2-CO-CH_2-CO-N_2H_3 + 2C_2H_5OH + 2H_2O \quad (A.9)$$

Preparation.   25.04 g of hydrazine hydrate (0.5 mol) is added dropwise to 40.05 g of diethyl malonate (0.25 mol) dissolved in 350 ml of absolute ethanol, in a 1000 ml round bottom flask fitted with a reflux condenser. The mixture is refluxed for about 5 h. The clear solution obtained is cooled and then concentrated on a water bath. White crystals are obtained which are filtered and dried (yield: 23.5 g, 70% and m.p. 152°C).[8]

A.2.7 *3-Methyl Pyrazole 5-One (3MP5O), $C_4H_6N_2O$*

3MP5O is prepared by the chemical reaction of 1 mol of ethyl acetoacetate with 1 mol of hydrazine hydrate. The chemical reaction may be written as follows:

$$CH_3-CO-CH_2-COOC_2H_5 + N_2H_4 \cdot H_2O$$
$$\longrightarrow C_4H_6N_2O + H_2O + C_2H_5OH \qquad (A.10)$$

Preparation.   3-Methyl pyrazole-5-one ($C_4H_6N_2O$) is prepared by drop wise addition of 1 mol ethyl acetoacetate to hydrazine hydrate (99.9%) cooled in an ice bath. The solution is kept overnight to crystallize. White (colorless) crystals form and are separated by filtration, dried and stored in a desiccator.[8]

A.3 USEFUL SUGGESTIONS

The combustion of homogeneous solutions containing redox mixtures generally yield crystalline solids directly. However, combustion of heterogeneous solutions containing redox mixtures along with solids like fumed silica or fire retarding substances like phosphates, yield amorphous products. These require further sintering in order to obtain single-phase crystalline product. The sintering step can be avoided by the addition of extra amount of energetic oxidizers like ammonium nitrate or perchlorate. These details have been mentioned in the preparation of mullite, cordierite, and NASICONS, etc.

Caution.   Hydrazine is a hazardous and explosive chemical. Whereas aqueous solution of hydrazine ($N_2H_4 \cdot H_2O$) is safer to handle. However, it is known to be carcinogenic and should be handled carefully taking precautions to avoid contact with body. All standard safety measures practiced in the chemical laboratory are to be followed.

References

1. Yamamura H, Watanabe A, Shirasaki S, Moriyoshi Y, Tanada M, Preparation of barium titanate by oxalate method in ethanol solution, *Ceram Int* **11**: 17–22, 1985.
2. Anuradha TV, Ranganathan S, Tanu Mimani, Patil KC, Combustion synthesis of nanostructured barium titanate, *Scr Mater* **44**: 2237–2241, 2001.
3. Mohr EB, Brezinski JJ, Audrieth LF, Carbohydrazide, in Bailer Jr JC (ed.), *Inorg Synth* **4**: 32–35, McGraw Hill, New York, 1953.
4. Gran G, The use of oxalyl dihydrazide in a new reaction for spectrophotometric microdetermination of copper, *Anal Chim Acta* **14**: 150–156, 1956.

5. Mashima M, The infrared absorption spectra of condensation products of formaldehyde with hydrazide, *Bull Chem Soc Jpn* **89**: 504–506, 1966.

6. Ainsworth C, Jones RG, Isomeric and nuclear substituted β-aminoethyl-1,2,4-triazoles, *J Am Chem Soc* **77**: 621–624, 1955.

7. Mizzoni RH, Spoerri PE, Synthesis in the pyridazine and 3,6-dichloropyridazine, *J Am Chem Soc* **73**: 1873–1874, 1951.

8. Ekambaram S, Combustion synthesis and properties of lamp phosphors, PhD Thesis, Indian Institute of Science, Banaglore, pp. 58–61, 1996.

Index